CHEMISTRY CONNECTIONS

CHEMISTRY CONNECTIONS

The Chemical Basis of Everyday Phenomena

Kerry K. Karukstis
Gerald R. Van Hecke

Department of Chemistry
Harvey Mudd College
Claremont, California

HARCOURT
ACADEMIC
PRESS

San Diego San Francisco New York Boston London Sydney Tokyo

Academic Press
a division of Harcourt Brace & Company
525 B Street, Suite 1900, San Diego, California 92101-4495, USA
http://www.apnet.com

Academic Press
24-28 Oval Road, London NW1 7DX, UK
http://www.hbuk.co.uk/ap/

Harcourt/Academic Press
200 Wheeler Road, Burlington, MA 01803
http://www.harcourt-ap.com

Library of Congress Catalog Card Number: 99-62310

International Standard Book Number: 0-12-400860-7

PRINTED IN THE UNITED STATES OF AMERICA
99 00 01 02 03 04 ML 9 8 7 6 5 4 3 2 1

The important thing is not to stop questioning.
—Albert Einstein

To our parents.

Contents

Preface   xiii
Conventions   xvii

Chapter 1

Connections to Medicine

How Does a Timed-Released Medicine Work?   1
**Why Are Iodine- and Barium-Based Fluids Used to Enhance
    CAT Scans?   5**
**Why Does the Patient Often Experience a Warm Flush after Taking
    an Iodinated Medium for a CAT Scan?   7**
Why Are General Anesthetics Administered as Gases?   9
How Do Sutures Dissolve?   15
What Is the Composition of an Artificial Hip?   18
What Is an Antidote for Cyanide Poisoning?   20

Chapter 2

Connections to Recreation

Why Do Light Sticks Glow?   23
**Why Should Pool Owners Add Chlorine to Their Swimming Pools
    in the Evening instead of the Morning?   28**
What Is Liquidmetal (in Liquidmetal Golf Clubs)?   30

Chapter 3

Connections to Food

**Why Won't a Glass of Water Relieve the Burning Sensation of
    Chili Peppers?   33**

Why Does Morton Salt Claim "When It Rains It Pours"?   35
What Causes Puff Pastry to Expand?   36
Why Is EDTA Added to Salad Dressings?   40
Why Does Soda Lose Its Carbonation as Its
    Temperature Rises?   44
Why Are Bubbles in a Soft Drink Spherical?   46
Why Are Metal or Glass Bowls Preferred for Whipping
    Egg Whites?   47
When a Flambé Dish Is Prepared, Why Is the Liqueur Heated Prior
    to Lighting the Flame?   49
Why Do Crystals Form in Wine over Time?   50

Chapter 4

Connections to Space

Why Is an Astronaut's Visor so Reflective?   53
What Is the Source of the Billows of White Smoke That Are
    Seen When the Booster Rockets Ignite upon Liftoff of the
    Space Shuttle?   55

Chapter 5

Connections to the Outdoors and the Environment

What Do Meteorologists Use to Seed Clouds?   57
Why Is the Aurora Borealis so Colorful?   60
Why Do Seashells Vary in Color?   62
Why Are Hydrangeas Pink When Grown in Arid Regions and Blue
    When Grown in Regions with a Heavy Rainfall?   63
Why Do Citrus Growers Spray Their Trees with Water to Protect
    Them from a Freeze?   66
Why Does a Mixture of Hydrogen Peroxide and Sodium
    Bicarbonate Deodorize a Dog That Has Been Sprayed by
    a Skunk?   68
If Natural Gas Is Odorless, Why Can We Detect the Odor of a
    Gas Leak?   70
What Does pH Stand for?   72

Chapter 6

Connections to the Office

How Do Scratch-and-Sniff Ads Work? How Does Carbonless
Copy Paper Work?   75

How Do "Repositional Self-Adhesive Paper Notepad Sheets"
(e.g., Post-it Notes) Work?   78
How Do Correction Fluids like Liquid Paper and
White-Out Work?   81
Why Does Disappearing Ink Disappear?   82
How Does an Automatic Fire Sprinkler Head Operate?   84

Chapter 7
Connections to the Household

What Is the Dark Spot on the Inside of a Light Bulb When It
Burns Out?   87
How Are Lightbulbs Frosted?   90
What Makes a Lava Lamp Work?   92
Why Shouldn't Ammonia Cleansers Be Mixed with Bleach?   95
Why Can Floor Waxes Be Removed Easily with
Ammonia Cleansers?   96
Why Do Manufacturers Recommend That Consumers Clean Their
Automatic Coffee Makers with Vinegar or Run Vinegar through
Their Steam Irons?   98
Why Do Homemade Recipes for Copper Cleaner Call
for Vinegar?   99
Why Does Baking Soda Extinguish a Fire?   101
Why Should One Never Attempt to Extinguish a Magnesium Fire
with either Water or a CO_2 Fire Extinguisher?   103
Why Is Hydrogen Peroxide Stored in Dark-Colored
Plastic Bottles?   104
How Do Furniture Polishes Repel Dust?   107
Why Does Soap Scum Form? (Or Why Are Phosphates Used
in Detergents?)   109
How Do Household Products such as Drano and Liquid-Plumr
Unclog Drains?   112
What Is Shatterproof Glass?   114
How Do Windshield Coatings Improve Visibility in
a Rainstorm?   118
What Causes the Pearlescent Appearances of Some Paints?   120
Why Are the Small Packs of Granules Labeled "Desiccants"
Enclosed in Bottles of Medication, New Shoes, and Electronics?
What Is the Identity of the Desiccants?   123
What Are the Red or Silver Liquids in Thermometers?   125
Why Does a Kitchen Gas Burner Glow Yellow When a Pot of
Boiling Water Overflows?   127

Why Does Superglue Stick to Almost Every Surface?   129
Why Are Ice Cubes Cloudy on the Inside?   131

Chapter 8

Connections to the Theater and the Arts

What Is the Origin of the Expression "In the Limelight"?   135
**Why Does a Flashbulb Develop a White Coating after
    a Flash?   138**
**Why Did Dorothy Have to Oil the Tin Woodman in *The Wonderful
    Wizard of Oz?*   139**
Why Do Old Paintings Discolor?   141
**What Caused the Destruction of the Organ Pipes in the Cathedrals
    of Northern Europe during the 19th Century?   144**
**How Do Fog Machines Create the Artificial Fog or Smoke Used in
    Theatrical Productions?   146**

Chapter 9

Connections to Currency and Gems

**Why Is It Incorrect to Call U.S. Currency
    "Paper Currency"?   151**
**What Is the Purpose of the Thread That Runs Vertically through
    the Clear Field on the Face Side of U.S. Currency?   154**
**Why Does the Denomination on the Lower Right-Hand Corner of
    New U.S. Currency Shift in Color from Green to Black When
    Viewed from Different Angles?   157**
**How Do Forensic Chemists Use Visible Stains to
    Trap Thieves?   161**
Why Is the Hope Diamond Blue?   164
Why Are Opals and Pearls Iridescent?   166

Chapter 10

Connections to Fabrics and Clothing

How Is a Fabric Made Water Repellent or Waterproof?   169
**Why Does a Bullet-Proof Vest Work? What Is the Composition of a
    Bullet-Proof Vest?   172**

Why Is a Cotton Towel More Effective in Cleaning Up Water Spills Than a Towel Made from a Synthetic Polyester Fiber? Why Do Cotton Fabrics Take Longer to Dry Than Synthetics? 177

What Is an Optical Brightener? 179

What Puts the "Blue" in Blue Jeans? 183

Chapter 11

Connections to Personal Care

Why Do Cosmetic Cold Creams Feel Cool When Applied to the Skin? 187

What Is an Alpha Hydroxy Acid? 188

What Is an "Alcohol-Free" Cosmetic? 192

What Causes the Cooling Sensation Found in Many Toothpastes and Breath Fresheners? 196

What Is the Difference between a Sunscreen and a Sunblock? 201

What Makes a No-Tears Shampoo? 206

What Causes the Fizz When an Antacid Tablet Is Added to Water? 209

What Is the Difference between Hard and Soft Contact Lenses? 212

Index 217

Preface

Chemistry Connections: The Chemical Basis of Everyday Phenomena highlights the fundamental role of chemical principles in governing our everyday experiences and observations. This collection of contemporary real-world examples of chemistry in action is written in a question-and-answer format with presentations in both lay and technical terms of the chemical principles underlying numerous familiar phenomena and topical curiosities. Introductory college chemistry students and educators as well as laypersons with an inquisitiveness about the world around them will find the book an informative introduction to the context of chemistry in their lives.

Assessment of the Need for This Book

United States Education Secretary Richard W. Riley recently commented on the results of a national assessment of scientific literacy among U.S. high school graduates[1]: "We are confronted by a paradox of the first order. We Americans are fascinated by technology. Yet, at the same time, Americans remain profoundly ignorant." Former National Science Foundation Director Neal Lane concurs: "I have become especially conscious of the discrepancy between the public's interest in, even fascination with, science and its limited knowledge about scientific concepts and issues".[2] He adds, "All scholarly fields—poetry or philosophy, architecture or agriculture—suffer from their separation from the public, although in the case of science, the separation may be more extreme. And yet science and the technology it spawns pervade the very structure of everyone's life. . . ."

Perhaps no scientific field is less understood and less appreciated by the public, and in particular by students, than chemistry. A general misunder-

[1] Woo, E. (1997, May 3). *Los Angeles Times.*
[2] *The Chronicle of Higher Education* (1996, December 6). p. A84.

standing of the nature of chemistry and even the meaning of the word *chemical* pervades our society. Students and the general public alike are further unaware of the broad scope of chemistry and the impact of the discipline on many fields. Conveying the importance and relevance of chemistry to our world is one of the greatest challenges facing chemists and chemical educators today.

The revitalization of chemistry education has received much recent attention and taken many forms. Modes of teaching, textbooks, laboratory instruction—all aspects of the chemistry curriculum—have undergone scrutiny for reform. A recent National Science Foundation report, *Shaping the Future: New Expectations for Undergraduate Education in Science, Mathematics, Engineering, and Technology,*[3] characterizes the nature of the most successful curricular and pedagogical improvements: "A simple precis is that these improvements are attempting to nurture a sense of wonder among students about the natural world, to maintain students' active curiosity about this world while equipping them with tools to explore it and to learn." Indeed, a recent survey of college chemistry courses revealed findings that indicated an increased emphasis toward the presentation of chemical principles reflecting "the more relevant chemistry of everyday living".[4] These initiatives are based on the development of a curriculum that engages a broad base of students and that provides students with a familiar context for chemical concepts, stimulating their desire to explore further.

Approach Used in This Text

By what mechanism do chemistry textbooks and monographs demonstrate the *relevance* of chemistry? Most general chemistry texts include short features highlighting real-world examples of the various chemical principles illustrated within each chapter. Some texts take a more revolutionary approach to promote the interest of students in chemistry by a textbook structure that focuses on key household products (e.g., food, apparel) and technologies or industries (e.g., health, communications, transportation) as a means of introducing the chemical principles of a standard college curriculum. Other trade books are broad overviews of the global impact of chemistry on society, usually written in nontechnical language.

In *Chemistry Connections* we have adopted a separate approach, collecting in one volume an assortment of provocative, topical questions that are raised by our everyday experiences and that are answered by the application of chemical principles. The design of the book makes it compatible with

---

[3] *Shaping the Future: New Expectations for Undergraduate Education in Science, Mathematics, Engineering, and Technology* (1996). Directorate for Education and Human Resources.
[4] Taft, H. L. (1996). *Journal of Chemical Education* **74,** 595–599

any general chemistry text for students and educators and suitable as an independent book for all individuals who are curious about the world around them. From the reader's viewpoint, the pertinence of chemistry to each question ranges from straightforward examples to more intriguing applications. We chose the question-and-answer format to provide a force motivating the reader to learn the chemical principles necessary for understanding everyday phenomena. Explanations are provided in both lay and technical terms—an initial description to satisfy the curious reader, followed by a more in-depth account to underscore the chemical nature of the phenomenon. We expect that readers will also quickly appreciate that an interplay of several chemical principles is often needed to explain fully a real-world observation, a realization too often overlooked by the beginning student or casual reader. Each question is indexed according to key principles or terms to provide teachers with the flexibility to select pertinent examples for class discussion. To furnish readers with related information for further exploration, we chose to focus on references to Web sites. With today's increasing access to the Internet, these selections may be more readily available than many hard-copy references. We recognize the transient nature of the World Wide Web, however, and encourage readers to use these sites as starting points for their own discovery of related electronic materials.

Acknowledgment

We acknowledge with gratitude the editorial assistance of Anna M. Hollifield (Harvey Mudd College Class of '99) during the preparation of this manuscript.

Kerry K. Karukstis
Gerald R. Van Hecke
Claremont, California

Conventions

The answers to many questions include chemical formulas shown in common drawing notation. For readers unfamiliar with such notation, the following examples should help in interpreting the figures. A line represents a chemical bond. The number of lines between two atoms represents the number of chemical bonds between the joined atoms. Unless stated otherwise, all lines join two carbon atoms. The junction of two lines implies the presence of a carbon atom, unless another atom is shown. Each carbon should have four "lines" drawn to it. Any missing line should be interpreted as a bond to a hydrogen atom. For example, a line that terminates should be viewed as a carbon bonded to three hydrogens and bonded to a fourth atom shown or implied. See the following drawings for examples.

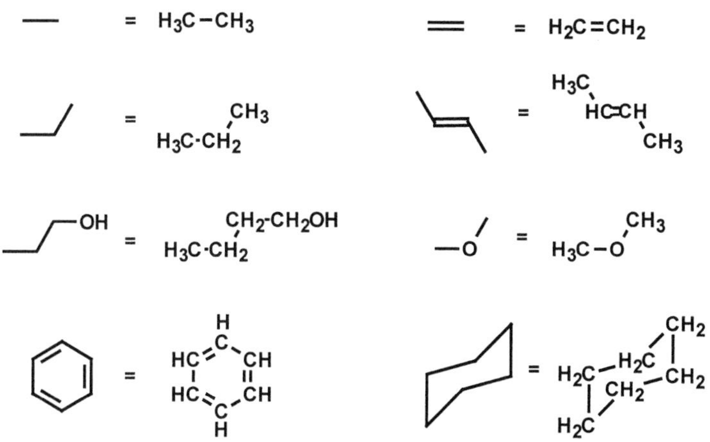

— = H_3C-CH_3

= = $H_2C=CH_2$

/ = $H_3C\text{-}CH_2$ CH_3

= H_3C $HC=CH$ CH_3

—OH = $H_3C\text{-}CH_2$ $CH_2\text{-}CH_2OH$

—O = H_3C-O CH_3

= $HC \overset{C}{\underset{C}{\overset{H}{}}} CH$ $HC \overset{}{\underset{H}{}} CH$

= H_2C-H_2C CH_2 CH_2-CH_2 H_2C

Chapter 1

Connections to Medicine

HOW DOES A TIMED-RELEASED MEDICINE WORK?

The formulation of time-released medicines is based on the specific response of polymeric coatings to their chemical environment. The chemical packaging of these medicines determines the precise conditions for effective control and sustained dosage of these drugs.

THE CHEMICAL ESSENCE

Physicians utilize a variety of protocols and therapies to heal and cure patients. Some medical treatments require the sustained application of a drug for maximum effectiveness. When hospitalized, a patient can receive this continued drug delivery through an intravenous (IV) unit. Some oral medications are chemically formulated to achieve this same effect. A medicine taken orally can often be gradually released in the body over a specified time interval by carefully designing the coating that encapsulates the medication. For example, the decongestant Contac contains numerous tiny beads of medicine that are covered by a water-soluble polymeric coating of varying thickness. The thicker the coating, the longer the time required for the coating to dissolve in water and the slower the release of the medicine. The claim of an effective "12-hour medicine" is based on the precise combination of beads of medicine with prescribed thicknesses of polymeric coating to sustain the controlled release for an extended period of time. Thin coatings obviously dissolve quickly, whereas thicker layers take longer (up

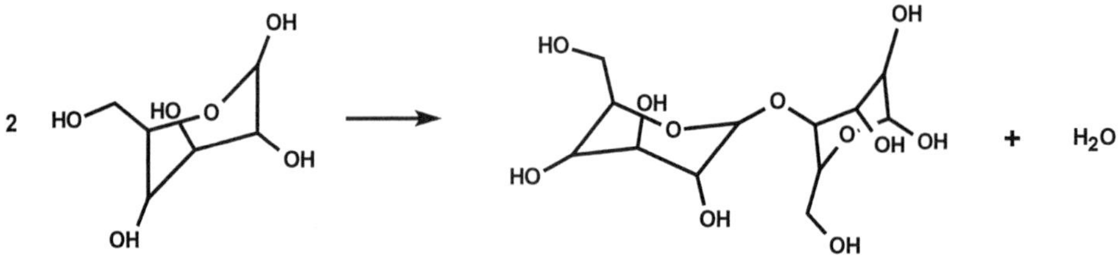

Figure 1 The formation of cellulose during a condensation polymerization reaction in which each new link of glucose monomers releases a water molecule.

to 12 hours, in this case). Timed-release formulations using "microencapsulation technology" are also used for agricultural purposes (e.g., fertilizers) and insect control (e.g., a six-month pest control).

THE CHEMICAL SPECIFICS

FMC designs one such water-soluble coating for timed-released medications—Aquacoat ECD, a 30% by weight aqueous dispersion of ethylcellulose. This polymer is used to coat drug-layered beads that are delivered using gelatin capsules for a pH-independent sustained release. Cellulose is a natural polymer containing repeating glucose units (monomers). Cellulose forms during a *condensation polymerization reaction* (Fig. 1) in which each new link of glucose monomers releases a water molecule. The $-OH$ and $-CH_2OH$ substituents on the glucose rings are replaced with $-OCH_2CH_3$ and $-CH_2OCH_2CH_3$ groups in ethylcellulose (Fig. 2).

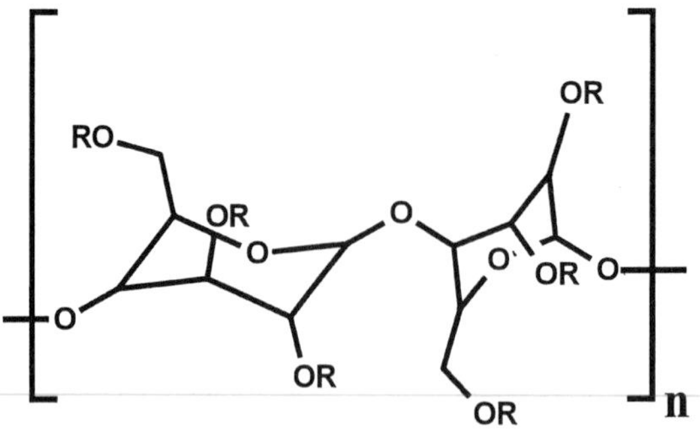

$R = CH_2CH_3$

Figure 2 The repeating unit of the polymer ethylcellulose.

Some extended-release preparations are designed with a coating that responds to the acidity of its environment. The polymeric coating of the medicine is formulated for stability during oral delivery and for eventual solubility at the intended organ. The contrasting acidic content of the stomach and the more basic environment of the intestines enable these formulations to function. For example, hydroxypropyl methylcellulose phthalate (HPMCP) (Fig. 3) is an *enteric* (i.e., solubilized in the intestinal tract) coating designed to protect acid-sensitive drugs from being destroyed by gastric acid in the stomach. In a more alkaline environment, deprotonation of the $-COOH$ carboxyl groups (to form $-COO^-$ carboxylate functionalities) is believed to enable the dissolution of the polymeric HPMCP, thereby releasing the encapsulated drug.[1] Enteric coating materials are specifically used to target timed-release medications to treat colon inflammations or other disorders of the digestive tract. In addition, enteric coatings are placed on aspirin caplets designed for the temporary relief of arthritic and rheumatic pain, muscle aches, joint pain, and back pain. The coating allows the caplet to pass through the stomach to the intestine before it dissolves to help prevent stomach irritation.

Polymer coatings responsive to temperature or moisture form the basis for medications delivered transdermally using patches on the skin or internally via inserts implanted in the body. Oral nitroglycerin (Fig. 4) tablets to prevent angina attacks or scopolamine (Fig. 5) to protect against motion sickness are two examples of drugs that can penetrate the skin and flow into the bloodstream at a rate dictated by a rate-controlling membrane in the patch. Hydrolysis of nitroglycerin leads to the formation of the reactive free radical nitric oxide, NO. NO activates guanylate cyclase to produce cyclic guanosine monophosphate (GMP); cyclic GMP decreases cellular

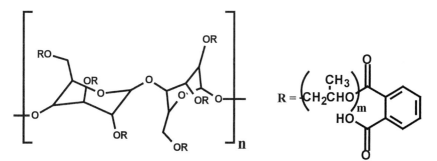

Figure 3 The repeating unit of the polymer hydroxypropyl methylcellulose phthalate (HPMCP).

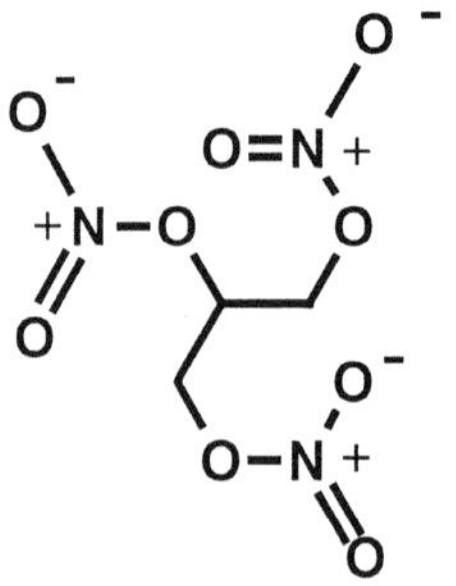

Figure 4   The chemical structure of nitroglycerin.

calcium levels, thereby causing dilation or expansion of the blood vessels to reduce myocardial oxygen demand.[2] Dilation of the arteries increases blood flow to the heart and relieves the chest pains that result from an insufficient supply of oxygen to the heart muscle. The sophistication of the technology of timed-release medications arises from the extensive structural control that the chemist has available for the design of polymeric coatings and matrices.

KEY TERMS

polymer; microencapsulation; enteric

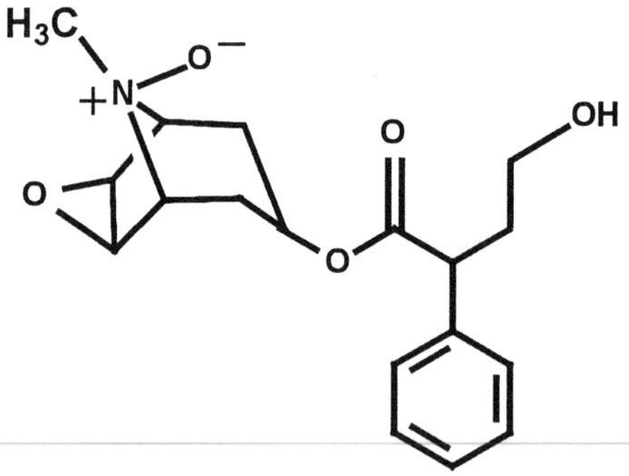

Figure 5   The chemical structure of scopolamine.

REFERENCES

1. Sanders, G. H. W., Booth, J., and Compton, R. G. (1997). Quantitative rate measurement of the hydroxide driven dissolution of an enteric drug coating using atomic force microscopy. *Langmuir* **13,** 3080–3083.
2. Ahlner, J., Andersson, R. G. G., Torfgard, K., and Axelsson, K. L. (1991). Organic nitrate esters: Clinical use and mechanisms of actions. *Pharmacol. Rev.* **43,** 351–423.

RELATED WEB SITES

"Aquacoat ECD," FMC Corporation, Pharmaceutical Division, Products, http://www. avicel.com/products/ec.html

"Drug Patches Catch On," Lalitha Gopinath, Global Bytes: *Chemistry in Britain,* September 1997, http://www.chemsoc.org/gateway/chembyte/cib/patches.htm

"Patches, Pumps and Timed Release: New Ways to Deliver Drugs," Marian Segal, U.S. Food and Drug Administration, http://www.fda.gov/bbs/topics/CONSUMER/CON00112.html

"Transdermal Drug Delivery Systems," PHA 3111 Pharmaceutics II, University of Florida, College of Pharmacy, Department of Pharmaceutics, http://prokai.cop.ufl.edu/tdds.htm

WHY ARE IODINE- AND BARIUM-BASED FLUIDS USED TO ENHANCE CAT SCANS?

The innovative and technological achievement of computerized axial tomography required a sophisticated understanding of physiology, mathematics, and radiology. A clear comprehension of chemical principles also contributes to the success of this diagnostic technique.

THE CHEMICAL ESSENCE

The technique of computerized axial tomography (CAT) assists medical examiners in viewing the internal organs of the body. The technique has been widely used since its development in the 1970s by the British electrical engineer Sir Godfrey Hounsfield and the South African-born U.S. physicist Allen Cormack. These scientists won the Nobel Prize for Physiology or Medicine in 1979 for their contributions to the development of this diagnostic technique.[1]

In a CAT scan cross-sectional, images are generated using x-rays directed through the body using a rotating tube. X-rays that are not absorbed by the body reach a radiation detector where the signals are integrated to produce an image that assesses the density of tissues at various locations.

X-rays are absorbed differentially—with denser objects such as bones absorbing extensively and soft tissues such as blood vessels absorbing relatively few x-rays. Thus bones appear as light areas on the image, soft tissues as dark regions.

What can one do to image specific soft tissues using x-rays? One must enhance the density of these regions to decrease the ability to transmit x-rays through the region. To aid in the creation of such images or CAT scans, *contrast media* are often ingested or injected into the body. Contrast-medium fluids that are opaque (i.e., not transparent) to x-rays are known as *radiopaque* media. These media highlight the areas of the body being scanned as a consequence of the inability of x-rays to penetrate these substances. Media containing barium or iodine meet these criteria. For imaging gastrointestinal tracts, barium-based media are orally ingested; for visualizing major blood vessels, iodinated contrast media are injected into the patient's veins using intravenous access.

THE CHEMICAL SPECIFICS

Why are barium- and iodine-based materials selected for contrast media? The production of x-ray images depends on the differences between the x-ray absorbing power of various tissues. This difference in absorbing power is called *contrast* and is directly dependent on tissue density. To artificially enhance the ability of a soft tissue to absorb x-rays, the density of that tissue must be increased. The absorption by targeted soft tissue of aqueous solutions of barium sulfate and iodized organic compounds provides this added density through the heavy metal barium and the heavy nonmetal iodine.

KEY TERMS

x-ray; contrast

REFERENCE

1. "Press release: The 1979 Nobel Prize in physiology or medicine," Nobelforsamlingen Karolinska Institutet, The Nobel Assembly at the Karolinska Institute, 11 October 1979, http://nobel.sdsc.edu/laureates/medicine-1979-press.html

RELATED WEB SITES

"Cardiac Imaging: History: A Timeline of Cardiac Imaging Modalities," Silver Platter, Physician's Home Page, 1996, http://cardiacimaging.silverplatter.com/side.htm

"Computed Tomography," Department of Diagnostic Radiology and Nuclear Medicine, Rush-Presbyterian-St. Luke's Medical Center, Chicago, http://www.rad.rpslmc.edu/~exams/ct/ct_dep.htm

WHY DOES THE PATIENT OFTEN EXPERIENCE A WARM FLUSH AFTER TAKING AN IODINATED MEDIUM FOR A CAT SCAN?

For the optimal viewing of soft tissues and organs by CAT scans, patients are often administered (either orally or intravenously) certain substances known as contrast media. Although these substances have the vital function of enhancing the differences in tissue density for the production of superior x-ray images, some minor side effects are often reported. Interestingly, it is the chemical nature of these contrast media that not only dictates their capacity for amplifying x-ray images but also triggers the reaction of patients to the media.

THE CHEMICAL ESSENCE

Patients often report a warm, flushed feeling as one of the side effects of injection of a contrast medium in preparation for a CAT scan. What causes this rise in body temperature? The answer lies in the ionic nature of the contrast medium and in the body's attempt to regulate the concentrations of all substances in the blood.

Introduction of a water-soluble ionic substance into the vascular system results in an increase in the number of particles in the bloodstream as the contrast substance dissolves. The body possesses several internal regulation systems and, when perturbed by an injection, attempts to restore the concentrations of substances in the blood to their "normal" or preinjection levels. To re-equilibrate the system, water from the cells of surrounding body tissue moves into the blood plasma through capillary membranes. This transfer of water is an example of *osmosis,* the diffusion of a solvent (water) through a semipermeable membrane (the blood vessels) into a more concentrated solution (the blood) to equalize the concentrations on both sides of the membrane. To accommodate the increase in water volume, the blood vessels must dilate or expand in size. An increase in the volume of circulating

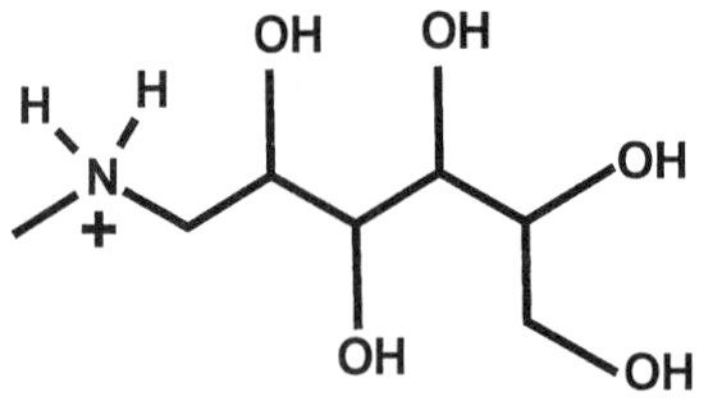

Figure 6   The chemical structure of the meglumine (*N*-methyl-D-glucamine) cation.

blood (due to the added water) causes an extra expenditure of energy by the body, producing the flushed or warm feeling. As this effect is a conse- quence of the process of osmosis, the effect is referred to as *osmotoxicity*. Side effects also associated with the increased blood volume and flow include excessive thirst and enhanced renal excretion.

THE CHEMICAL SPECIFICS

Some of the ionic iodinated substances used in CAT scans include amido- trizoates (salts of diatrizoic acid) and iothalamates (salts of iothalamic acid).[1] The typical cations are sodium ion or meglumine (*N*-methyl-D- glucamine) ion (Fig. 6) and diatrizoate (Fig. 7) and iothalamate (Fig. 8) are the common anions. These water-soluble substances have a relatively high iodine content relative to the organic portion of the substance. For example, iotalamic acid has a molecular formula of $C_{11}H_9I_3N_2O_4$ and a corresponding molecular weight of 613.92 g mol^{-1}. On a percent weight

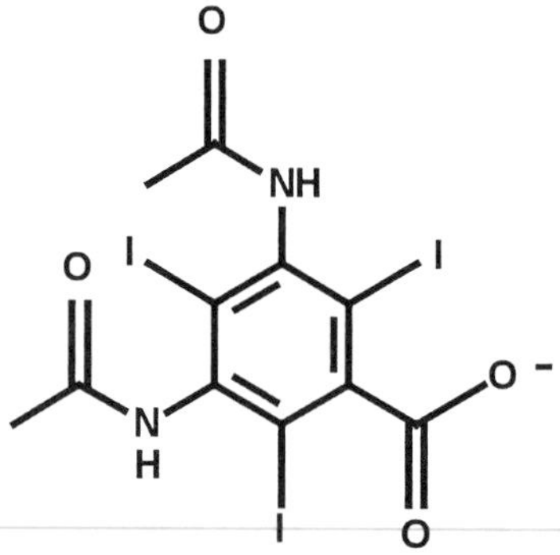

Figure 7   The chemical structure of the diatrizoate anion.

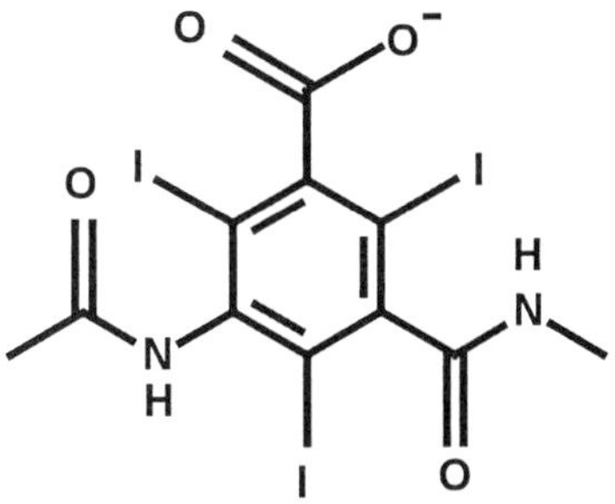

Figure 8 The chemical structure of the iothalamate anion.

basis, iotalamic acid contains 62.0% iodine. A high iodine concentration is desirable to enhance the absorption of x-rays while only introducing a small amount of the contrast agent into the bloodstream and body tissue.

KEY TERMS

osmosis; dissolution

REFERENCE

1. "OutSource: Contrast Media Part I," Wilbur L. Reddinger, Jr., http://www.t2star.com/ct_1/contrast1.html

RELATED WEB SITES

"HYPAQUE ORAL, Sanofi Winthrop, Diatrizoate Sodium, Contrast Medium," RxMed, The Website for Family Physicians, http://www.rxmed.com/monographs/hypaque.html
"Mallinckrodt Iothalamate Sodium, Radiopaque Medium," RxMed, Website for Family Physicians, http://www.rxmed.com/monographs/conray2.html
"The South African Medicines Formulary: CONTRAST MEDIA," http://www.uct.ac.za/depts/pha/samf/v08html.htm

WHY ARE GENERAL ANESTHETICS ADMINISTERED AS GASES?

Many surgical procedures employ the rapid, safe, and well-controlled application of gaseous anesthetics, often mixed with oxygen. These drugs are admin-

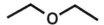

Figure 9   The chemical structure of ether (diethyl ether).

istered to obtain various stages of consciousness, muscular relaxation, and sensory stimulation. One of the characteristics of an ideal anesthetic agent is the rapid induction of these sensations using gaseous or easily volatilized inhalants. The chemical structure of an anesthetic dictates its physical state.

THE CHEMICAL ESSENCE

General anesthetics are drugs that are administered to depress the brain's sensory response. The effectiveness of anesthetic gases depends on their ability to directly dissolve in the bloodstream and circulate to the brain. Thus, the ability of a gas to dissolve in liquids, particularly the aqueous solution that comprises blood, is critical to the function of the anesthetic. How does the drug reach the neural tissue in the brain? An equilibrium of the inhaled gas with the lung tissue first occurs to transfer the drug through the walls of the alveoli within the lungs to the arterial blood. The drug then circulates to all tissues of the body, including the brain. A subsequent equilibrium of the anesthetic between the blood and the brain tissue determines the patient's level of unconsciousness.

THE CHEMICAL SPECIFICS

Some of the general anesthetics include ether (diethyl ether) (Fig. 9), chloroform (Fig. 10), nitrous oxide (Fig. 11), ketamine hydrochloride (Fig. 12), and halogenated hydrocarbons such as ethyl chloride (Fig. 13), trichloroethylene (Trilene) (Fig. 14), halothane (Fluothane or 2-bromo-2-chloro-1,1,1-trifluoroethane) (Fig. 15), methoxyflurane (Penthrane, Metofane)

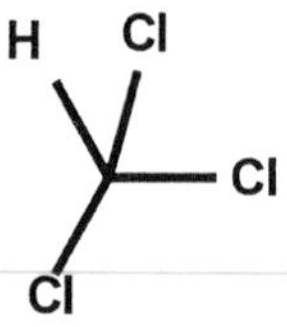

Figure 10   The chemical structure of chloroform.

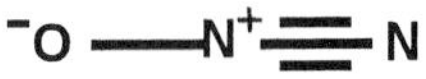

$$^-\text{O} \longrightarrow \text{N}^+ \equiv \text{N}$$

Figure 11 The chemical structure of nitrous oxide.

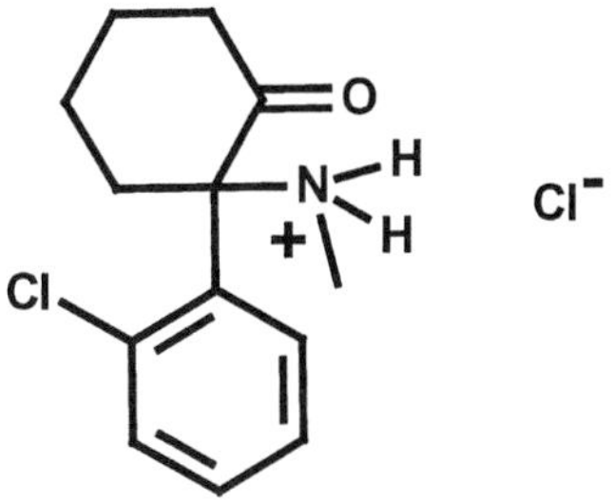

Figure 12 The chemical structure of ketamine hydrochloride.

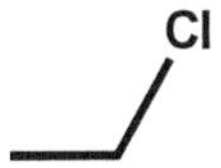

Figure 13 The chemical structure of ethyl chloride.

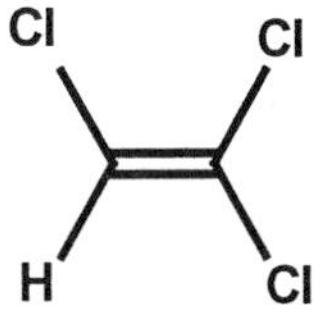

Figure 14 The chemical structure of trichloroethylene (Trilene).

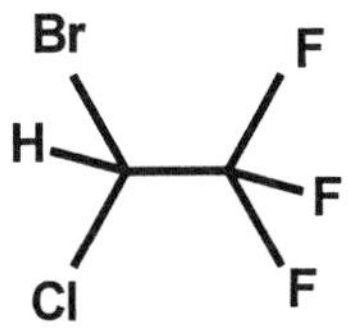

Figure 15 The chemical structure of halothane (Fluothane or 2-bromo-2-chloro-1,1, 1-trifluoroethane).

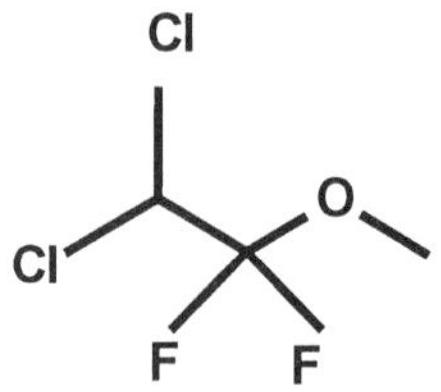

Figure 16   The chemical structure of methoxyflurane (Penthrane or Metofane).

(Fig. 16), enflurane (Enthrane or 2-chloro-1,1,2-trifluoroethyl difluoromethyl ether) (Fig. 17), isoflurane (Forane, Aerrane, or 1-chloro-
2,2,2,trifluoroethyl difluoromethyl ether) (Fig. 18), Sevoflurane (Ultane or
1,1,1,2,2,2-hexafluoroethyl fluoromethyl ether) (Fig. 19), and desflurane
(Suprane or 1,2,2,2-tetrafluoroethyl difluoromethyl ether) (Fig. 20).[1] Except
for nitrous oxide, all of these substances are liquids at room temperature.
Thus, the anesthetic agent must be vaporized upon administering and is
usually delivered in combination with oxygen gas.

Despite the fact that the anesthetics listed are liquids at room temperature, they all are characterized by a relative ease of *volatility* (i.e., tendency
to vaporize or become a gas). The strength of intermolecular forces (i.e.,
forces between molecules) dictate the facility with which the liquid to
gas phase transition can be accomplished. Weak dipole-dipole interactions
predominate in these generally polar compounds. No hydrogen bonds can
form between molecules, as no hydrogen atoms are present that are bonded
to electronegative atoms such as oxygen, nitrogen, or fluorine. Thus, to
change the physical state of the anesthetics from liquid to gas, only weak
intermolecular forces need to be overcome.

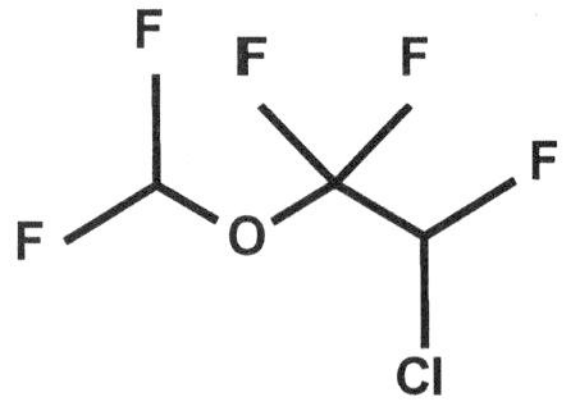

Figure 17   The chemical structure of enflurane (Enthrane or 2-chloro-1,1,2-trifluoroethyl
difluoromethyl ether).

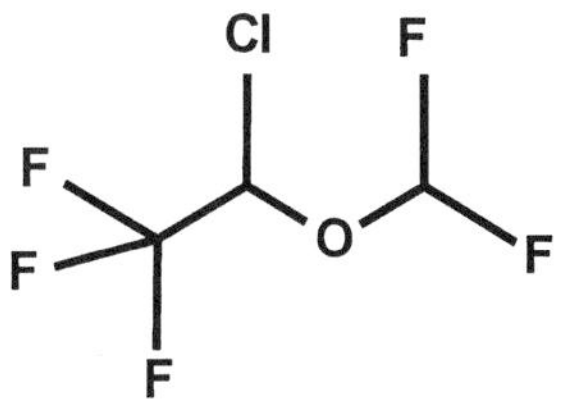

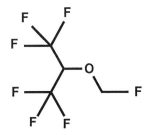

Figure 18 The chemical structure of isoflurane (Forane, Aerrane, or 1-chloro-2,2,2-trifluoro-ethyl difluoromethyl ether).

The ease of vaporization of these substances is evident in the low magnitudes of their boiling points. For example, ether has a normal boiling point of 35°C, isoflurane 48.5°C, halothane 50°C, enflurane 56.5°C, and trichloroethylene 87°C. The high volatility of these drugs is also reflected in their high vapor pressures at or near room temperature. For example, at 20°C sevoflurane has a vapor pressure of 157 mm Hg,[2] enflurane 175 mm Hg,[3] isoflurane 238 mm Hg,[3] and desflurane 669 mm Hg.[3] For comparison, the vapor pressure of water at 25°C is 23.8 mm Hg.

The ability of the anesthetic agent to function is related to the partial pressure of the drug in the brain. Two major factors dictate the concentration of anesthetic agent in the neural tissue: (1) the pressure gradients from lung alveoli to the brain (i.e., inhaled gas → alveoli → bloodstream → brain and (2) the lipid solubility of the drug that enables it to pass between the blood-brain barrier to the central nervous system. The distribution of anesthetic throughout the entire body may be viewed as an equilibration

Figure 19 The chemical structure of Sevoflurane (Ultane or 1,1,1,2,2,2-hexafluorethyl fluoromethyl ether).

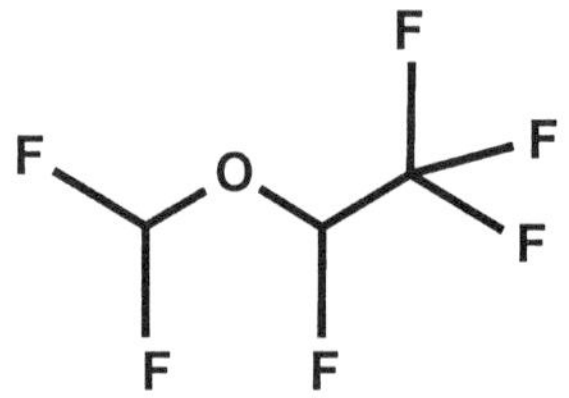

Figure 20 The chemical structure of desflurane (Suprane or 1,2,2,2-tetrafluoroethyl difluor-omethyl ether).

process (Fig. 21), with tissues characterized by high blood flows reaching equilibration faster than muscle and fat.[4] Nevertheless, an anesthetic that is excessively soluble in blood will not partition substantially into brain and other tissues.

The anesthetic properties of nitrous oxide and diethyl ether have been known since the 1840s. Zeneca Pharmaceuticals introduced the first modern inhalation anesthetic Fluothane in 1957.[5] Methoxyfluorane followed in 1960, enflurane in 1973, isoflurane in 1981, desflurane by Anaquest (Liberty Corner, New Jersey) in 1992, and sevoflurane by Abbott Laboratories in 1995.[6]

KEY TERMS

vapor pressure; equilibrium; vaporization; boiling point; volatility; intermolecular forces

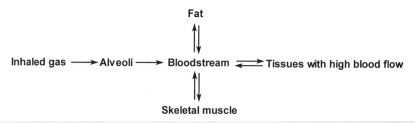

Figure 21 An overall description of the distribution of anesthetic throughout the body as an ensemble of equilibrium processes.

REFERENCES

1. "Chemistry of General Anesthetics," *PHA 422—Neurology Pharmacotherapeutics—Medicinal Chemistry Tutorial,* Patrick M. Woster, Ph.D., Department of Pharmaceutical Sciences, College of Pharmacy and Allied Health Professions, Wayne State University, http://wizard.pharm.wayne.edu/medchem/ganest.html
2. "New Anesthetic Requires New Vaporizers for Safety," William Clayton Petty, M.D., Department of Anesthesiology, Uniformed Services University of the Health Sciences, Bethesda, MD, http://gasnet.med.yale.edu/apsf/newsletter/1996/winter/apsf-new_anes.html
3. "Anesthetic Vapor Pressure Information," Scott Medical Products, http://www.scottgas.com/medical/anestech.html
4. "Veterinary Anesthesia: Inhalation Anesthesia," Western College of Veterinary Medicine, University of Saskatchewan, http://www.usask.ca/wcvm/anes/anesman/anes07.htm
5. "Zeneca Pharmaceuticals Therapeutic Areas: Central Nervous System," Zeneca Pharmaceuticals, http://www.zeneca.com/zpharma/cns.htm
6. "FDA Advisory Panel Recommends Approval of Abbott's New Anesthesia Drug," Abbott Laboratories OnLine, http://www.abbott.com/news/1995news/pr011895.htm

RELATED WEB SITES

"Economic value of Desflurane in comparison to Isoflurane in low flow anesthesia," I. Kuhn, M.D., and H. Wissing, M.D., Department of Anesthesiology, University Hospital Frankfurt/Main, Germany, Theodor Stern Kai, D-60590 Frankfurt/Main, Germany, http://gasnet.med.yale.edu/esia/1998/may/economic.html

"Nitrous Oxide from Discovery to Now," *Dental Digest,* http://www.dentaldigest.com/nitrous/nittext.html

"Using Volatile Anaesthetic Agents," Dr. P. Fenton, *Update in Anaesthesia, Practical Procedures,* Issue 5, 1995, http://users.ox.ac.uk/~ndainfo/wfsa/html/u05/u05_007.htm

HOW DO SUTURES DISSOLVE?

History tells us that catgut suture had its origin around A.D. 150 in the time of the Greek physician Galen, who built his reputation by treating wounded gladiators.[1,2] What chemical materials do modern surgeons use for sutures to ensure biocompatibility, tensile strength, and dissolution by the body's natural action?

THE CHEMICAL ESSENCE

Have you ever wondered how sutures could possibly hold a wound together long enough to promote healing *and* manage to dissolve over time? And what about the added need for flexibility so that the surgeon can

Figure 22   The synthesis of poly(glycolic acid) (PGA) from the dimer of glycolic acid.

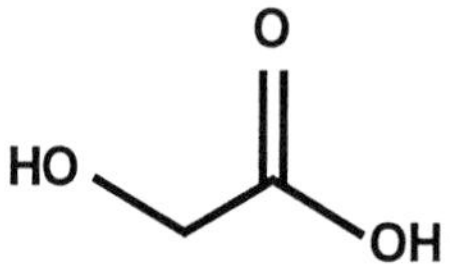

Figure 23   The chemical structure of glycolic acid.

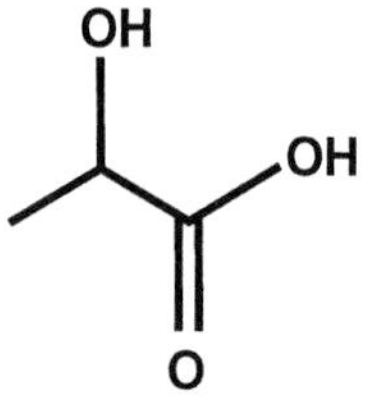

Figure 24   The chemical structure of lactic acid.

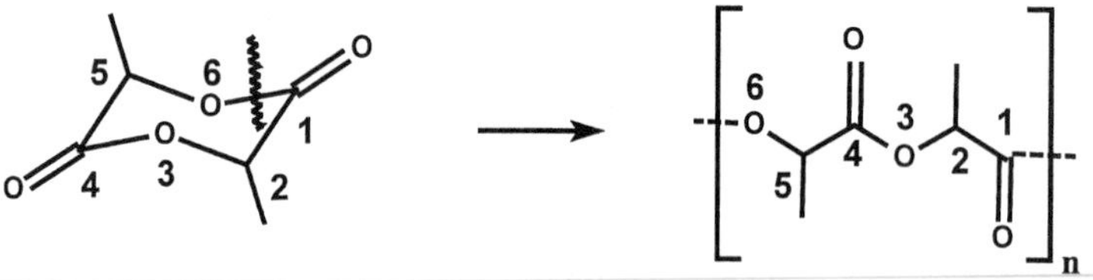

Figure 25   The synthesis of poly(lactic acid) (PLA) by a ring opening polymerization of the cyclic diester of lactic acid (lactide).

tie a knot that holds? *Synthetic biodegradable polymers* have the desirable mechanical and chemical properties to perform all of these functions. Some biodegradable polymers degrade by *hydrolysis* (i.e., by a reaction with water), others degrade *enzymatically* (i.e., via a reaction with an enzyme in the body). The products of the degradation process are harmless to the body.

THE CHEMICAL SPECIFICS

Common biodegradable polymers for medical devices are constructed from synthetic linear aliphatic polyesters. One material commonly used for internal sutures is poly(glycolic acid) (PGA). PGA is synthesized from the dimer of glycolic acid (Fig. 22).[1]

PGA degrades by hydrolysis to produce carbon dioxide and glycolic acid (Fig. 23), which is either excreted or enzymatically converted to other metabolized species. Lactic acid (Fig. 24) has also been polymerized into poly(lactic acid) (PLA) and developed into commercial sutures. PLA is prepared from the cyclic diester of lactic acid (lactide) by ring opening polymerization (Fig. 25).[1]

The lactic acid generated by the hydrolytic degradation of PLA becomes incorporated into one of the normal metabolic cycles of the body and is excreted as carbon dioxide and water. Copolymers of glycolic acid and lactic acid are also commonly used as biodegradable sutures to balance the greater strength of PGA and the slower degradation of PLA.

KEY TERMS

polymers; hydrolysis

REFERENCES

1. "Notes: Bioadsorbable Polymers," William B. Gleason, University of Minnesota, http://www.courses.ahc.umn.edu/medical-school/BMEn/5001/
2. "Biomedical Applications of Textiles: Sutures," Alice Baker, Will Fowler, Sophie Guevel, Allen Smith, North Carolina State University, http://www.bae.ncsu.edu/bae/research/blanchard/www/465/textbook/biomaterials/projects/textiles/fowler/project1.html

RELATED WEB SITES

"A Brief History of the Origins of Sutures," The Veterinarian's Sutures Guide, Dr. R. A. Henderson, http://www.vetmed.auburn.edu/~hendera/guide/guide2.htm

"Materials," Dr. David Harrison, School of Education, North East Wales Institute of Higher
    Education, Wrexham, North Wales, http://www.newi.ac.uk/buckleyc/materials.htm#Poly-
    mers and Plastics
"Applications: Medical Devices," Birmingham Polymers, Inc., http://www.bpi-sbs.com/bpi/

WHAT IS THE COMPOSITION OF AN ARTIFICIAL HIP?

The success of hip replacements has been greatly advanced by major developments in the biomaterials for orthopedic devices. What aspects of chemistry must be considered in designing an effective artificial hip?

THE CHEMICAL ESSENCE

Biomaterials are synthetic and naturally occurring materials that are foreign to the body but are used to replace a diseased organ or tissue or to augment or assist a partially functioning organ or tissue. Cardiovascular, orthopedic, and dental applications are some of the most common areas in which biomaterials are employed.

Successful applications of materials in medicine have been experienced in the area of joint replacements, particularly artificial hips. As a joint replacement, an artificial hip must provide structural support as well as smooth functioning. Furthermore, the biomaterial used for such an orthopedic application must be inert, have long-term mechanical and biostability, exhibit biocompatibility with nearby tissue, and have comparable mechanical strength to the attached bone to minimize stress. Modern artificial hips are complex devices to ensure these features.

Designed as a cup and ball joint, the artificial hip or prosthesis consists of two parts: the stem or femoral component and the socket or acetabular component. Inserted into the thighbone or femur, the metal stem is composed of titanium alloy or a cobalt-chromium-molybdenum alloy with a metal or ceramic ball-shaped head. Anchored to the hip bone, the acetabular component is a metal hemispherical shell with a plastic lining to act as a bearing for the femoral head. Figure 26 illustrates the two structural components of the artificial hip joint. Both the stem and the socket may be cemented into place with a special epoxy-type cement to bond the metal component to the bone. Alternatively, newer cementless surgical procedures use a porous coating of a calcium-phosphate-based ceramic to promote bone growth into porous surfaces. Hip implants of the future will replace the metal components with fiber composites that will more evenly match the stiffness of bone.

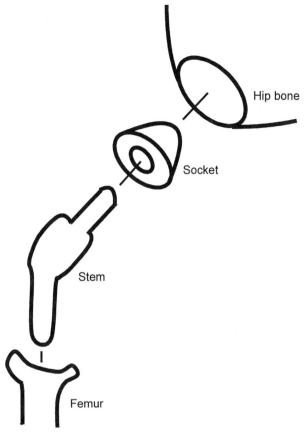

Figure 26 A schematic diagram of the two components of an artificial hip: the stem or femoral component and the socket or acetabular component.

The Chemical Specifics

The composition of bone is primarily a two-phase system of collagen, an organic substance, and calcium hydroxyapatite [$Ca_{10}(PO_4)_6(OH)_2$], an inorganic calcium phosphate mineral. Variations in the mineral-collagen ratio lead to changes in physical properties. Increased proportions of collagen tend ultimately to greater flexibility, whereas a higher level of mineral leads to increased brittleness. At the present time, no synthetic material mimics perfectly the mechanical properties of bone. Metals such as titanium alloys or cobalt-chromium-molybdenum alloys have the necessary strength and fracture resistance but have high degrees of stiffness. Ceramics such as hydroxyapatite generally possess the hardness and compressibility of bone but not the fracture resistance. Polymeric materials such as polymethylmethacrylate and polyethylene have improved performance due to low stiffness and reasonable fracture resistance but generally lack strength. In addition to mechanical properties, the biocompatibility and chemical degradation properties of biomaterials for implants must also be addressed. For example, the selection of metals such as titanium ensures a high degree of corrosion resistance in physiological media (i.e., water with dissolved oxygen, hydronium ions, chloride ions, etc.). Titanium derives its corrosion resistance by forming a protective oxide surface coating of TiO_2. Although a material with properties identical to bone has not yet been devised, the example of the design of biomaterials for total hip replacement illustrates the chemical and material advances that ultimately improve the quality of human life.

KEY TERMS

biomaterial; alloy

RELATED WEB SITES

"Case Study of Materials Selection for Total Hip Replacement," Materials Department, Queen Mary and Westfield College, University of London, http://www.materials.qmw. ac.uk/implant/

"Hydroxyapatite Crystals/Prosthetic Coating," Wheeless' Textbook of Orthopaedics, C. R. Wheeless, M.D., 1996, http://www.medmedia.com/o2/49.htm

"Metal-on-Metal: A Potential Alternative to Polyethylene," *Orthopedics Today,* http://www. slackinc.com/bone/ortoday/199605/metal.htm

"New Wedge-Shaped Hip Stem to Enter Market," *Orthopedics Today,* 1996, http://www. slackinc.com/bone/ortoday/199607/fetto.htm

"A Patient's Guide to Artificial Hip Replacement," Randale Sechrest, M.D., Medical Multimedia Group, http://www.sechrest.com/mmg/thr

"Titanium," Wheeless' Textbook of Orthopaedics, C. R. Wheeless, M.D., 1996, http:// www.medmedia.com/o16/69.htm

WHAT IS AN ANTIDOTE FOR CYANIDE POISONING?

The high toxicity of cyanide-containing substances requires the immediate application of a medical antidote. What simple chemical principles dictate both the toxicity of this species and the efficacy of suitable antidotes?

THE CHEMICAL ESSENCE

Cyanide (actually cyanide ion) is a particularly toxic substance as a consequence of its strong affinity for the metal-containing enzymes responsible for providing energy for cell respiration. In particular, cyanide forms stable complexes with the metal iron (i.e, with ions of iron—ferric ions, Fe^{3+}) in the respiratory enzyme cytochrome oxidase located in the mitochondria of cells. When the ferric ions are complexed with cyanide, the functioning of cytochrome oxidase as an essential catalyst for oxygen utilization in cells is inhibited. Normal cell functions cease when cell respiration is impaired, leading to cell mortality. Antidotes for cyanide poisoning offer cyanide ion the possibility of forming even more stable complexes with a different metal ion, leaving the ferric ions free to function properly within cytochrome oxidase.

THE CHEMICAL SPECIFICS

One antidote for acute cyanide poisoning used by toxicologists and physicians is the product known as Kelocyanor©. One ampule of Kelocyanor© contains 300 mg of the cobalt salt dicobalt edetate or dicobalt EDTA, Co_2EDTA (Fig. 27).[1] EDTA (ethylenediaminetetraacetic acid) is a multidentate ion with up to six lone pairs of electrons that are capable of forming six *coordinate covalent bonds* with metal ions. Four of these coordinating sites involve lone pairs of electrons on an oxygen atom in the acetate ligand, and two coordinating sites involve lone pairs of electrons on each of the

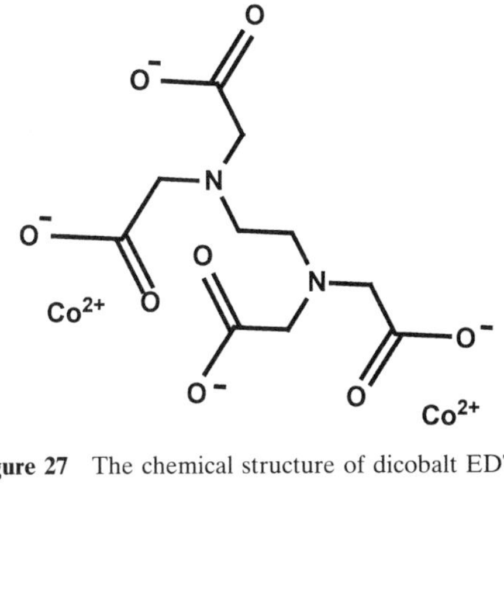

Figure 27   The chemical structure of dicobalt EDTA.

nitrogen atoms. In dicobalt edetate, a neutral substance, each Co^{2+} ion is coordinated to two acetate ligands. The formation constant for this complex is large:[2]

$$2 \ Co^{2+} + 4 \ EDTA \rightarrow Co_2EDTA \ \ K_{formation} = 2.0 \times 10^{16}$$

The basis for the toxicological activity of this substance is the reaction of cobalt ion with cyanide ion to form a relatively nontoxic and stable ion complex. The hexacyanocobaltate ion contains a Co^{2+} central metal ion with six cyanide ions as ligands. This *coordination complex* involves six *coordinate covalent bonds* whereby each cyanide ion supplies a pair of electrons to form each covalent bond with the central cobalt ion. The formation constant for the hexacyanocobaltate ion is even larger than for dicobalt EDTA,[3] and thus the cobalt ion preferentially exchanges an EDTA ligand for six cyano ligands:

$$Co^{2+}(aq) + 6 \ CN^- \ (aq) \rightarrow Co(CN)_6^{4-} \ (aq) \ \ K_{formation} = 1.2 \times 10^{19}$$

The exchange of EDTA for the CN^- ion reduces the concentration of cyanide ion in the body, making the cobalt ion an effective scavenger of toxic cyanide ions.

KEY TERMS

complex formation; complex ion; coordinate covalent bond; coordination complex

REFERENCES

1. "Antidotes: Kelocyanor," S.E.R.B. Laboratories, Paris, http://www.serb-labo.com/b3_b4_antidote.html
2. Skoog, D. A., West, D. M., and Holler, F. J. (1996). *Analytical chemistry,* 6th ed., p. 242. Saunders College Publishing, Philadelphia.
3. Peters, D. G., Hayes, J. M., and Hieftje, G. M. (1974).*Chemical Separations and Measurement,* p. A.12. W. B. Saunders Company, Philadelphia.

RELATED WEB SITE

"Cyanide Poisoning," S.E.R.B. Laboratories, Paris, http://www.serb-labo.com/c1_cyanide_poison.html

Chapter 2

Connections to Recreation

WHY DO LIGHT STICKS GLOW?

Few of us recognize that the fascinating glow of a light stick at a carnival or fair arises from a complex series of chemical reactions. The chemist's palette—a variety of ingredients with specific chemical structures—produces the rainbow of colors seen in these novelties.

THE CHEMICAL ESSENCE

Light sticks and other glow-in-the-dark products involve a phenomenon similar to the processes that lead to a firefly's glow or a lightning strike. This phenomenon is called *chemiluminescence*—the process whereby chemical energy produced by a chemical reaction is transformed into light energy, a different form of energy. In a firefly the light-producing chemical reaction is triggered by a catalyst known as an enzyme. In a lightning storm, an electrical discharge in the atmosphere sets off a sequence of reactions that leads to the flash of light in the sky. In a light stick, when you follow a manufacturer's instructions to bend, snap, and shake the stick, you initiate a chemical reaction by mixing the substances contained within separate compartments in the plastic tube. Energy generated during the course of the reaction is accepted by a dye molecule contained within the light stick and then released in the form of colored light with no accompanying heat.

The CYALUME light sticks, marketed by Omniglow Corporation, have found an extensive range of applications beyond the initial toy and novelty market. The intense yet cool light provided by a chemical light stick is ideal for emergency lighting, traffic control, and hazard identification. Numerous sports products, including golf balls, footballs, hockey pucks, badminton birdies, and wiffle balls, use replaceable light sticks to create innovative nighttime activities. In virtually every amusement park open past sunset, we can find varieties of glow-in-the-dark colored necklaces, bracelets, and bands. Military covert, nighttime, and emergency operations have been enhanced with the use of visible and infrared chemiluminescent products. The United States and Allied forces during Operation Desert Storm employed light sticks for underwater operations, nighttime personnel and ship-to-ship identification, hazard demarcation, and color-coding of units, vehicles, and equipment. Commercial fishing fleets have found that deep-sea fish such as tuna and swordfish are attracted to the light emitted by light sticks when the fish swim in surface waters at night. In particular, the long-line fishing industry utilizes light sticks to illuminate their monofilament long lines, often extending in length up to 80 miles with more than 3000 hooks. With the variations in color, intensity, and duration of light emission possible, numerous applications in other commercial industries are likely.

THE CHEMICAL SPECIFICS

The color, intensity, and duration of light that we observe depend on the exact composition of a light stick. Let's explore how the chemical composition affects these properties of the light stick.

The amount of energy produced by the chemical reaction between the original components in the light stick dictates the color of the glow. Certain chemical reactions generate large amounts of energy and give rise to short-wavelength light (i.e., energy $\propto 1/\text{wavelength}$) that is blue or violet in color. Other reactions are somewhat less energetic and give rise to red and orange glows at the longer wavelength end of the visible spectrum.

The efficiency of the conversion of energy from the chemical reaction into light is one factor that affects the intensity of the light emission. The more efficient the conversion, the brighter the light stick's glow. In addition to a characteristic energy output, each chemical reaction has its own efficiency.

The intensity of a light stick is also sensitive to temperature. An increase in temperature accelerates the rate of the chemical reaction, increasing the intensity or amount of light that can be generated from the dye molecule in a given time period. By the same token, you can prolong the life of a light stick by storing it in the freezer. At lower temperatures, the chemical reaction slows down, producing less energy per unit time and yielding less intense light that persists for a longer period of time. By making light intensity measurements at a variety of temperatures, adherence to the Arrhenius law can be demonstrated (i.e., intensity $\propto$ rate $\propto e^{-1/T}$).

The duration of the light stick's glow is dictated by the amount of chemical reactants contained in the tubing. At any temperature, once all of the original chemicals contained within the light stick are consumed by the reaction, chemiluminescence is no longer possible, and the light stick fades to darkness.

Let's look at a general chemical reaction for the chemiluminescence of a light stick in some detail. Hydrogen peroxide, H_2O_2, is the primary ingredient contained in an aqueous solution within an inner thin-glass ampule within the lightstick. By bending the plastic tubing of the light stick, the thin vial of H_2O_2 is broken, mixing H_2O_2 with a surrounding solution of a phenyl oxalate ester $((COOC_6H_5)_2)$ and a fluorescent dye. The ester and peroxide react in a series of several steps to generate a highly energetic C_2O_4 intermediate as in Figure 1.[1]

Variations in the substitution on the phenyl rings of the oxalate ester primarily affect the yield of C_2O_4 intermediate. The energetic nature of the intermediate presumably arises from the significant strain imposed by the four-membered ring. Recall that the ideal bond angle around a planar sp^2-hybridized carbon is 120°, not 90°. The chemielectronic step—the step in which the *chemical* energy of the intermediate is converted into *electronic* energy in the fluorescent dye—may be written as in Figure 2.

The release of energy to the dye molecule or fluorescer is driven by the conformational instability of the C_2O_4 intermediate (the flat highly strained C_2O_4 prefers to be two linear CO_2 molecules). The sensitized fluorescer, denoted fluorescer*, returns to the ground state via the emission of light:

$$\text{fluorescer*} \rightarrow \text{fluorescer} + h\nu$$

Emission spanning the visible and near infrared wavelengths has been obtained through the choice of fluorescent dye. For example, 9,10-

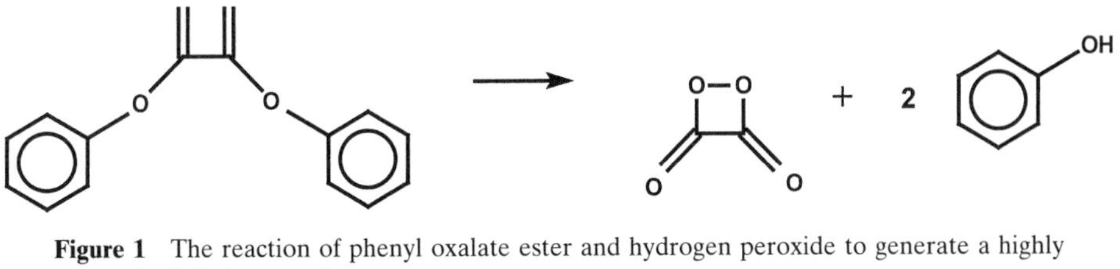

Figure 1 The reaction of phenyl oxalate ester and hydrogen peroxide to generate a highly energetic C_2O_4 intermediate.

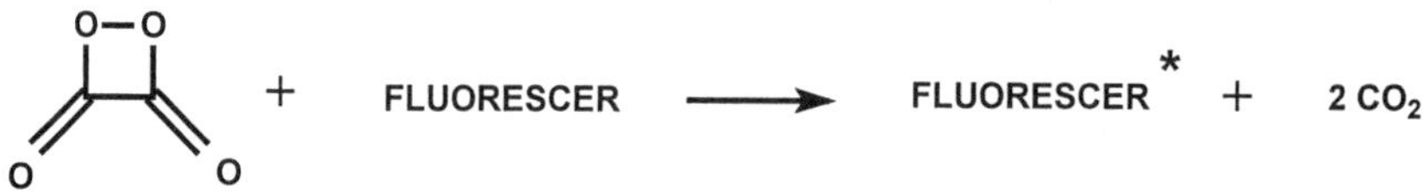

$$\text{O-O} \quad + \quad \text{FLUORESCER} \quad \longrightarrow \quad \text{FLUORESCER}^* \quad + \quad 2\,CO_2$$

Figure 2 A chemielectronic step (i.e., a step in which the chemical energy of an intermediate is converted into electronic energy in a fluorescent dye). Here the C_2O_4 intermediate releases energy as it dissociates to two carbon dioxide molecules. The energy is transferred to the accompanying fluorescent dye to generate an excited state of the dye.

diphenylanthracene (Fig. 3) generates blue light; 9,10-bis(phenylethynyl) anthracene (Fig. 4) yields yellow-green emission with maximum output at 486 nm; rubrene (5,6,11,12-tetraphenylnaphthacene; Fig. 5) emits orange-yellow light at 550 nm; violanthrone (Fig. 6) emits orange light at 630 nm; 16,17-(1,2-ethylenedioxy)violanthrone (Fig. 7, R...R $= -OCH_2CH_2O-$) gives rise to a red glow at 680 nm; and 16,17-dihexyloxyviolanthrone (Fig. 7, R $= -OC_6H_{13}$) provides infrared luminescence at 725 nm.

KEY TERMS

chemiluminescence; chemical reaction rate; Arrhenius law; wavelength and energy of light

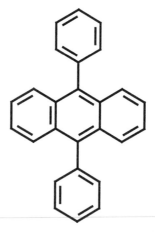

Figure 3 The chemical structure of 9,10-diphenylanthracene.

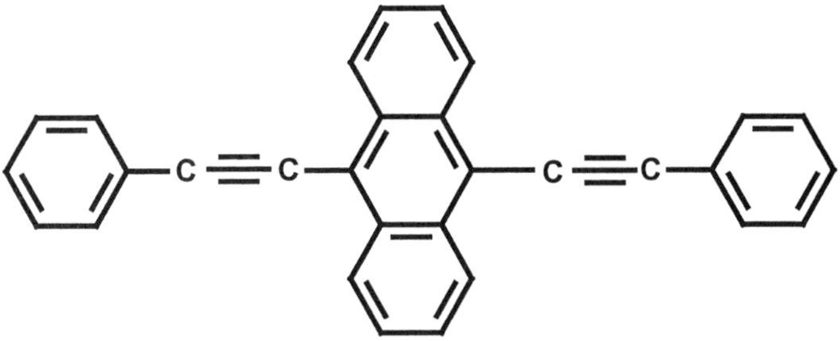

Figure 4   The chemical structure of 9,10-bis-phenylethynyl)anthracene.

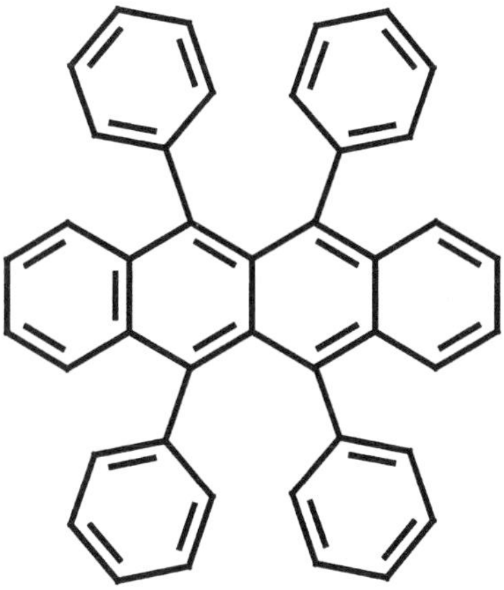

Figure 5   The chemical structure of rubrene (5,6,11,12-tetraphenylnaphthacene).

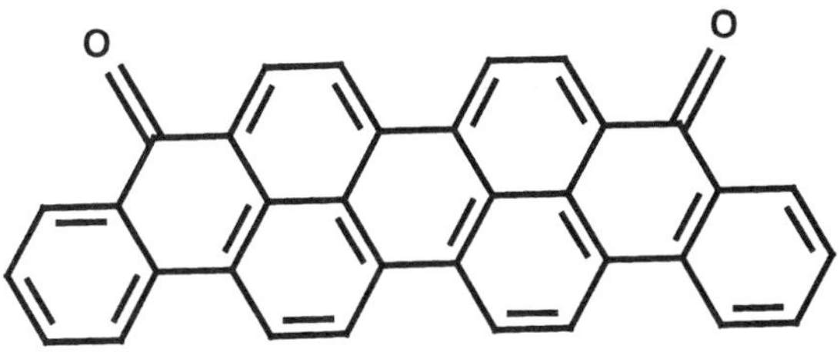

Figure 6   The chemical structure of violanthrone.

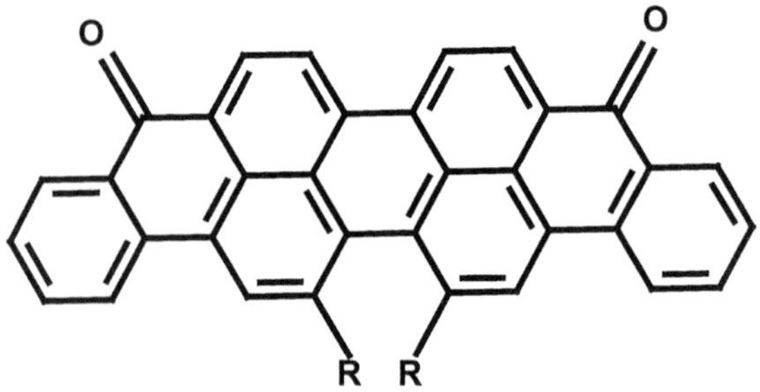

Figure 7 The chemical structure of 16,17-(1,2-ethylenedioxy)violanthrone with R..R = $-OCH_2CH_2O-$ and 16,17-dihexyloxyviolanthrone with each R = $-OC_6H_{13}$.

REFERENCE

1. Shakhashiri, B. Z. (1983). Lightsticks. *In: Chemical demonstrations: A handbook for teachers of chemistry,* Vol. 1, pp. 146–152. The University of Wisconsin Press, Madison.

RELATED WEB SITES

"The Chemiluminescence Home Page," Dr. Thomas G. Chasteen, Sam Houston State University, http://www.shsu.edu/~chm_tgc/chemilumdir/chemiluminescence2. html
"Lightstick Chemistry," Moravian College, Chemistry Department, http://www.cs.moravian.edu/chemistry/lightstick/l_st_scheme.html
"Lightstick Spectra," Moravian College, Chemistry Department, http://www.cs.moravian.edu/chemistry/lightstick/l_st_spec.html
Omniglow Corporation, http://www.omniglow.com/

WHY SHOULD POOL OWNERS ADD CHLORINE TO THEIR SWIMMING POOLS IN THE EVENING INSTEAD OF THE MORNING?

One swimming pool dealer[1] makes the following claim: "Timing is everything—if you want an easy pool! Adding chlorine to your swimming pool in the evening, instead of the morning, can cut your chemical costs in half." What valuable swimming pool chemistry is the basis for this assertion?

THE CHEMICAL ESSENCE

Chlorine, or more commonly a substance containing hypochlorite ion, is added to pools as a disinfectant. However, sunlight rapidly destroys hypochlorite, drastically reducing the effectiveness of the sanitizer. Hence, the effectiveness of the disinfectant is maximized when added in the evening hours.

THE CHEMICAL SPECIFICS

Pool chlorine, generally added in the form of calcium hypochlorite $(Ca(OCl)_2)$ or sodium hypochlorite (NaOCl) readily ionizes in water to yield the hypochlorite ion, OCl^-:

$$Ca(OCl)_2 \text{ (aq)} \rightarrow Ca^{2+} \text{ (aq)} + 2\ OCl^- \text{ (aq)}$$

$$NaOCl \text{ (aq)} \rightarrow Na^+ \text{ (aq)} + OCl^- \text{ (aq)}$$

The hypochlorite ion, a weak base, can react further with water to generate hypochlorous acid (HOCl) and hydroxide ions (OH^-) that raise the pH of the pool water:

$$OCl^- \text{ (aq)} + H_2O \rightarrow HOCl \text{ (aq)} + OH^- \text{ (aq)}$$

Alternatively, upon absorption of ultraviolet light, a photochemical reaction can occur—that is, a reaction triggered by the energy provided from light. This reaction destroys the hypochlorite ion and produces chloride ion and oxygen gas, which escapes from the pool:

$$OCl^- \rightarrow Cl^- + 1/2\ O_2 \text{ (g)}$$

This reaction is also an oxidation-reduction process whereby the oxygen atom is oxidized from the -2 oxidation state to the zero oxidation state as the chlorine atom is reduced from the $+1$ to -1 oxidation state. As diatomic oxygen is an effective disinfectant, pool owners should avoid the loss of O_2 via the decomposition of the hypochlorite ion. Adding hypochlorite-containing disinfectant in the evening hours reduces the loss of the ion from photochemical decomposition.

KEY TERMS

oxidation-reduction; photochemical reaction

REFERENCE

1. "Swimming Pool Secret #6," Pool Solutions, WaterCare, Inc., http://www.poolsolutions.com/tip06.html

RELATED WEB SITES

"Pool Water Chemistry: Technical Details," Virtual Pool and Spa Store, http://www.poolandspa.com/page371.htm

WHAT IS LIQUIDMETAL (IN LIQUIDMETAL GOLF CLUBS)?

"Breakthrough Technology That Will Change the Game Forever."[1] *This is the claim of the manufacturers of an innovative combination of metals that is revolutionizing the golf industry. Inserted into the club heads of putters, irons, and drivers, this material provides golfers with more solid shots, with greater energy upon impact, and with enhanced accuracy upon trajectory than conventional clubs. What is the chemistry of this unique material?*

THE CHEMICAL ESSENCE

Liquidmetal is a patented five-metal alloy made by combining nickel, zirconium, titanium, copper, and beryllium. Liquidmetal is the discovery of two scientists from the California Institute of Technology in Pasadena. Dr. Atakan Peker, senior scientist, and Dr. William L. Johnson, Mettler Professor of Materials Science at Caltech, invented Liquidmetal in the spring of 1992. Liquidmetal Golf, the Laguna Niguel, California, company that markets golf clubs made with the alloy, claims that the structure of this innovative combination of metals produces a stronger, lighter, more resilient material that transfers more energy to a golf ball upon impact.

THE CHEMICAL SPECIFICS

Liquidmetal is a bulk amorphous or non-crystalline alloy that contains approximately two-thirds zirconium as well as nickel, titanium, copper, and beryllium. In the materials world, this type of alloy is often described as a liquid metal, a metallic glass, or a "glassy metal." The origin of this terminology lies in the microstructure that the alloy adopts as it is cooled from a molten state. Conventional metals and alloys form crystalline structures with defined three-dimensional arrangements of atoms or molecules. By cooling the molten form of Liquidmetal rapidly, the metallic atoms in Liquidmetal do not arrange in an ordered manner but in random fashion. This *amorphous* state is known as a *glassy* state, as glassy substances lack the rigid regularity of a crystalline solid. As the molecules of liquids also are not locked into definitive locations, the alloy thus has properties characteristic of liquids, hence the name Liquidmetal.

Why does this amorphous structure improve the performance of Liquidmetal? Crystalline materials are actually composed of many small crystals that are held together in a polycrystalline structure. Each single crystal containing a regular arrangement of atoms is known as a grain. Although grains are packed together tightly, the shapes of grains are irregular and lead to *grain boundaries* when the crystals come into contact. The grain boundaries of conventional crystalline metals lead to microscopic gaps, and pockets are potential sites of weakness. The amorphous structure of Liquidmetal means that there are no grains and no grain boundaries to cause defects in the alloy's structure.

The composition of an alloy dictates its particular properties. Liquidmetal has an extremely high-tensile strength of 1900 MNm^{-2}, twice that of titanium or stainless steel.[1,2] The manufacturer also claims a high degree of resiliency for Liquidmetal, two to three times higher in resistance to deformation than conventional metals.[1] Combined with its low density[2] of 6.1g cm^{-3} (lower than stainless steel but higher than titanium[1]), the alloy has a very high strength-to-weight ratio compared to other materials. The application in golf club heads takes full advantage of this strength-to-weight ratio. The claim that more energy is transferred to the ball on impact results from the metal absorbing less energy than conventional materials on impact. More power leads to greater distances per shot. The alloy also possesses exceptional vibrational damping qualities, reducing the shock on impact and providing a soft but solid feel. The alloy also exhibits a high degree of hardness, an important feature to reduce wear.

KEY TERMS

alloy; amorphous; crystalline; glass

REFERENCES

1. "Properties of Liquidmetal Alloy," Liquidmetal Golf, http://www.liquidmetalgolf.com/facts/facts.htm
2. "Liquidmetal Swings into Action," *Materials World,* September 1998, http://www.materials.co.uk/mwldweb/sept98/feat-cr.htm

Chapter 3

Connections to Food

WHY WON'T A GLASS OF WATER RELIEVE THE BURNING SENSATION OF CHILI PEPPERS?

Don't reach for that glass of water to cool the effect of spicy chili peppers! A few rules of chemistry will suggest a better remedy.

THE CHEMICAL ESSENCE

Why does water "cool" the burning sensation of some "hot" (i.e., spicy) foods and not others? The cooling action arises as water acts to *dilute* the concentration of substances responsible for the burning sensation, thereby reducing their effect. To make a more dilute solution, water must be capable of dissolving the pungent ingredient. The common expression "like dissolves like" is the operational key. For any two substances, the closer their chemical structure, the more similar their intermolecular forces, that is, their attractive interactions on the molecular level. Strong intermolecular interactions lead to enhanced *solubility* (i.e., ability to dissolve). Water will cool the effect of those spicy ingredients that it can dissolve.

Substances that dissolve readily in water exhibit similar intermolecular forces to those displayed by water. In other words, compounds soluble in water are polar substances, often with the capacity to hydrogen bond to water. From your own experiences, you already recognize substances that dissolve in water and those that do not. For example, salt (ionic sodium chloride) dissolves readily in water, whereas oils such as butter and margarine have a limited solubility that can be enhanced by an increase in temperature. Many "hot" spices are generally non-polar substances that dissolve only partially in water. Examples of spices with molecular structures that limit the degree of dissolution in water include zingerone, a constituent of ginger; piperine, the active component of white and black pepper; and capsaicin, the pungent ingredient in red and green chili peppers as well as paprika.

The chemical structure of these spices renders these substances practically insoluble in water and other aqueous solutions but freely soluble in alcohols, oils, and fats (i.e., organic solvents). Hence, a drink of cold water will only temporarily relieve the burning sensation induced by the action of these ingredients on the pain-detecting nerve endings in the mouth. Alternatively, the pungency of ginger or pepper is better alleviated with a fat-based accompaniment such as sour cream or a beverage containing some alcohol.

THE CHEMICAL SPECIFICS

The pungent components of chili peppers belong to a class of substances known as *capsaicinoids*. The most pungent and most common substance in this family is *capsaicin* (Fig. 1) (N-[(4- hydroxy-3-methoxyphenyl)-methyl]-8-methyl-6-nonenamide). Other members of this family include *dihydrocapsaicin* (Fig. 2), *nodihydrocapsaicin* (dihydrocapsaicin with a $(CH_2)_5$ linkage instead of $(CH_2)_6$), *homocapsaicin* (capsaicin with a $(CH_2)_5$ unit instead of $(CH_2)_4$), and *homodihydrocapsaicin* (dihydrocapsaicin with $(CH_2)_7$ instead of $(CH_2)_6$). The capsaicinoid content varies with the pepper. In bell peppers capsaicin and dihydrocapsaicin are present in about a $1:1$ ratio, whereas in Tabasco peppers the ratio is closer to $2:1$.

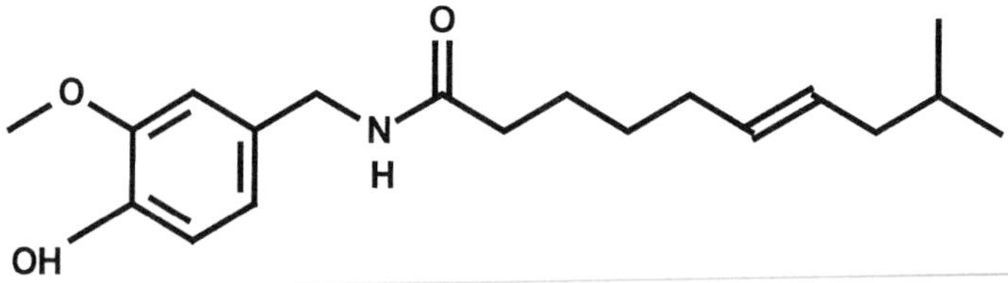

Figure 1 The chemical structure of capsaicin N-[(4-hydroxy-3-methoxyphenyl)methyl]-8-methyl-6-noneamide.

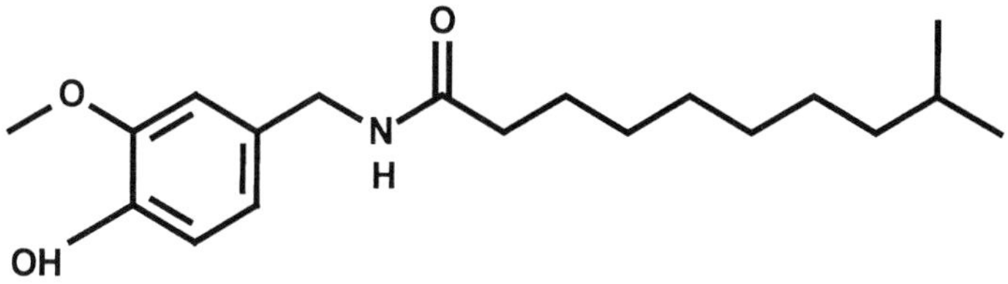

Figure 2　The chemical structure of dihydrocapsaicin.

KEY TERMS

solubility; hydrogen bonding

RELATED WEB SITES

"Scientific Information on Chile Peppers," Mike Bowers, University of California, Davis, http://neptune.netimages.com/~chile/science.html

WHY DOES MORTON SALT CLAIM "WHEN IT RAINS IT POURS"?

The Morton Salt girl carrying her familiar umbrella during a rain shower is a well-recognized caricature for today's consumers. Together with her slogan, "When It Rains It Pours," this image was introduced by the Morton Salt Company to promote a chemical innovation in this important household product.

The Chemical Essence

In 1911 the Morton Salt Company introduced the slogan "When It Rains It Pours" with their familiar little girl with an umbrella to promote, in their words, "a packaging innovation"[1] that allowed salt to pour easily even in humid conditions.[1] In fact, their innovation is actually an additional ingredient or "free-flowing agent"[1] that prevents salt from absorbing water and caking. Such an ingredient has the capacity to absorb many times its weight in water without itself dissolving, allowing the table salt to pour freely from its container.

THE CHEMICAL SPECIFICS

What was the first free-flowing agent added to Morton Salt in 1911? Magnesium carbonate, $MgCO_3$, was the first anti-caking agent used by the Morton Salt Company.[2] The extra component present in Morton Salt packages today is calcium silicate, $CaSiO_3$. This white powder has the incredible ability to absorb liquids and still remain a free-flowing powder. In general, calcium silicate absorbs 1 to 2.5 times its weight of liquids. For water, its total absorption powder is estimated as 600%, that is, it absorbs 600 times its weight of water.[3] Morton International indicates that the amount of calcium silicate added to a package of table salt is "less than one-half percent" (by weight).[4] In addition to adding calcium silicate to table salt, this anticaking agent is also included in formulations of baking powder.

KEY TERMS

anti-caking agent

REFERENCES

1. Morton Salt, Morton International, http://www.mortonintl.com/salt/oversalt.htm
2. General FAQs, Morton Salt, Morton, International, http://www.mortonintl.com/salt/faqs/fagmfaqs.htm
3. *The Merck Index,* 10th ed. (1983), p. 234. Merck & Co., Inc., Rahway, NJ.
4. Morton Salt, Morton International, http://www.mortonintl.com/salt/faqs/fatsfaqs.htm

RELATED WEB SITES

"Food Additives," Dr. Anthony Lopez, Department of Food Science and Technology, Virginia Polytechnic Institute and State University, http://www.fst.vt.edu/cfast/foodad.html
Morton Salt, Morton International, http://www.mortonintl.com/salt/oversalt.htm
The Salt Institute, http://www.saltinstitute.org

WHAT CAUSES PUFF PASTRY TO EXPAND?

Who hasn't been tantalized and impressed by the delectable, eye-catching specialties prepared with puff pastry: chocolate napoleons, Beef Wellington, baked Brie, pate en croute . . . yum! Although the secret of these culinary

*delights may seem more of an art than a science, professional pastry chefs
know the chemistry necessary to create the flaky, tender pastry for these deli-
cacies.*

THE CHEMICAL ESSENCE

The baking process creates the flaky layers in puff pastries (*pate feuil-
letee*) used for rich napoleon desserts and chicken *vol-au-vents* and generates
the air pockets in chou pastes that are filled to create cream puffs and
eclairs. Heat alone will not achieve the desired effect. Substances known
as *leavening agents* cause doughs and batters to expand through the release
of gases in these baking mixtures. Common leavening agents include substances such as yeast, baking soda, and baking powder. But a leavening
effect can also be achieved by entrapping air in the batter through vigorous
beating ("air leavening," as in angel food cakes and sponge cakes) and by
the vaporization of volatile liquids due to heat from an oven. When the
volatile liquid is water, the process is referred to as *steam leavening.* Water
vapor pressure is fairly insignificant at room temperature but rises substantially as the boiling point of water is approached. The volume expansion
of steam creates the puffed nature of the pastry or the interior cavities for
cream puffs and eclairs.

THE CHEMICAL SPECIFICS

What are the gases produced from the heating of leavening agents?
When the leavening agent is baking soda or sodium bicarbonate ($NaHCO_3$),
the gas carbon dioxide (CO_2) is released when the baking soda combines
with an acidic ingredient in the recipe:

$$NaHCO_3 \text{ (s)} + H^+ \text{ (aq)} \rightarrow Na^+ \text{ (aq)} + H_2O \text{ (l)} + CO_2 \text{ (g)}$$

Common acidic ingredients include vinegar, lemon juice, sour milk, but-
termilk, yogurt, tart fruits, and cream of tartar. Commercial bakeries often
use ammonium bicarbonate or ammonium carbonate as a leavening agent.
The gas-producing reaction with ammonium bicarbonate actually generates
both carbon dioxide gas and ammonia gas:

$$NH_4HCO_3 \text{ (s)} \rightarrow NH_3 \text{ (g)} + H_2O \text{ (l)} + CO_2 \text{ (g)}$$

The distinctive aroma of ammonia is often apparent in bakeries but not in
the final product. Bakers' yeast performs its leavening function by ferment-
ing such sugars as glucose, fructose, maltose, and sucrose. The principal

products of the fermentation process are carbon dioxide gas and ethanol, an important component of the aroma of freshly baked bread. The fermentation of the sugar, glucose—an example of a decomposition reaction—is given by the following equation in Figure 3.

The key to the action of steam leavening is the temperature-dependent vapor pressure of water. The vapor pressure of a liquid at a given temperature is the pressure of the gas in equilibrium with the liquid. The liquid-gas equilibrium is characterized by equal rates of vaporization and condensation at the molecular level (a "dynamic" equilibrium) and a constant or "equilibrium" vapor pressure at the macroscopic level. Two key factors affect a liquid's vapor pressure—the intermolecular forces exerted between molecules and the liquid's temperature. The stronger the intermolecular forces, the less likely the liquid is to escape to the gas phase, thus keeping the vapor pressure low. The higher the temperature, the greater the motion of the molecules and the more likely the molecules are to overcome the existing intermolecular forces, increasing the vapor pressure. The dependence of the vapor pressure of water on temperature is illustrated in Figure 4. The curve on the left is generated by plotting vapor pressure directly as a function of temperature. On the right, a linear relationship is obtained by graphing ln P as a function of the reciprocal of temperature. This relationship is expressed mathematically in the form of the Clausius-Clapeyron equation:

$$\ln P = \frac{-\Delta H_{vap}}{R}\left(\frac{1}{T}\right) + \frac{\Delta S_{vap}}{R}$$

The slope of the line allows for the determination of the enthalpy of vaporization of water, ΔH_{vap}, and the y-intercept yields the entropy of vaporization, ΔS_{vap}. As both the enthalpy and the entropy of water increase as the phase change *liquid → vapor* occurs, the slope and y-intercept of the Clausius-Clapeyron equation are negative and positive, respec-

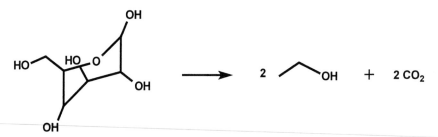

Figure 3.   The fermentation of sucrose to yield ethanol and water (a decomposition reaction).

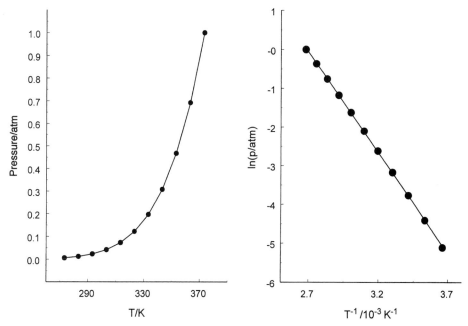

Figure 4 A graph of the vapor pressure of liquid water as a function of temperature (on the left), and a graph of the natural logarithm of the vapor pressure of water as a function of the reciprocal of temperature.

tively. At 373 K these thermodynamic quantities have values of ΔH_{vap} = 40.657 kJ mol[-11] and ΔS_{vap} = 109.0 J K^{-1} mol[-12].

The leavening action due to water vapor or steam arises from the increased amount of water vapor that forms as pastry temperatures initially rise in the oven and then from the increased volume of the water vapor as temperatures continue to rise to the desired baking temperature. Let's look at this volume expansion in quantitative detail. First of all, how do the volume of one mole of liquid water and one mole of liquid vapor at 373 K and 1 atm compare? The volume occupied by one mole or 18.02 g of liquid water at these conditions is equal to the quantity *weight/density* or (18.02 g mol^{-1})/(0.95840 g cm^{-3}) = 18.80 cm^3 = 18.80 mL.[1] Using the ideal gas law, $PV = nRT$, the corresponding amount of water vapor occupies (1 mol)(0.08206 L atm mol^{-1} K^{-1})(373 K)/(1 atm) = 30.6 L, a factor of 1630 times larger!

The ideal gas law further reveals that the greater the number of vapor molecules at a given temperature, the greater the volume exhibited by the gaseous phase. So as batter in an oven increases in temperature to 373 K, liquid water molecules undergo a phase transition to water gas or vapor, increasing the number of gas molecules and hence the volume of space occupied by the now gaseous water. At most baking temperatures, all water present will be in the vapor phase. As the temperature of dough rises, the greater the volume exhibited by the water vapor, again in accord with the ideal gas law. A baking temperature of 400°F (204°C or 477 K) induces a 1.3-fold increase in water vapor volume (477 K/373 K) compared with 100°C (212°F or 373 K) conditions. Although water vapor can escape from the pastry, the volume expansion that occurs prior to "escape" produces the leavening action that leads to flaky pastry layers or even large void cavities.

KEY TERMS

vapor pressure; leavening agent; ideal gas law; enthalpy of vaporization; entropy of vaporization; Clausius-Clapeyron equation

REFERENCES

1. Lide, D. R. (Ed.). (1993). *Handbook of chemistry and physics,* 74th ed., p. 6-10. CRC Press, Boca Raton, FL.
2. Atkins, P. (1994). *Physical chemistry,* 5th ed., p. C17. W. H. Freeman, New York.

RELATED WEB SITES

"Baking Powder and Baking Soda: What's the Difference?" Inquisitive Cook Online, http://www.inquisitivecook.com/articles/faceoff.shtml
"Baking Soda vs. Baking Powder," http://www.users.interport.net/~sue/food/bakgsoda.html
"The Cook's Thesaurus: Leavening Agents," Lori Alden, http://www.northcoast.com/~alden/Leaven.html

WHY IS EDTA ADDED TO SALAD DRESSINGS?

An inspection of the ingredients in many sandwich spreads, mayonnaises, margarines, and salad dressings reveals the abbreviation "EDTA." Even "real mayonnaise" has this important ingredient. The chemical structure of this substance helps this additive perform its important function as a preservative.

THE CHEMICAL ESSENCE

The presence of unwanted metal ions in foods and beverages can often be traced to their presence in soils and in the machinery used for harvesting and processing of food. In particular, contamination by even trace amounts of copper, iron, or nickel is especially undesirable because these metals are known to catalyze the reaction of oxygen with unsaturated fats in foods, leading to undesirable color changes and rancidity. Food spoilage can be retarded or eliminated by including certain food additives known as *sequestrants*. These additives, of which EDTA is an example, form tightly bound complexes with the trace metals, preventing the catalytic function of the metal ions and diminishing the degradation of the food item. Salad dressings, mayonnaise, margarine, processed fruits and vegetables, canned shellfish, and soft drinks are common food items with added EDTA.

THE CHEMICAL SPECIFICS

The sodium and calcium salts of EDTA (ethylenediaminetetraacetic acid, Fig. 5) are common sequestrants in food products. A three-dimensional representation of EDTA is shown in Figure 6. The EDTA ion is an especially effective sequestrant, forming up to six *coordinate covalent bonds* with a metal ion. These bonds are so named because a lone pair of electrons on a single atom serves as the source of the shared electrons in the bond between the metal ion and EDTA. The two nitrogen atoms in the amino groups and the four oxygen atoms in the carboxyl groups of EDTA are

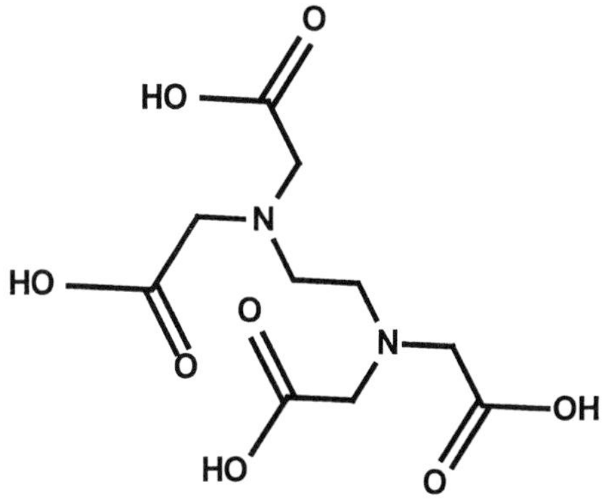

Figure 5   The chemical structure of ethylenediaminetetraacetic acid (EDTA).

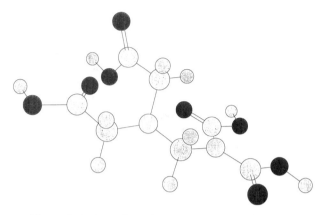

Figure 6   A three-dimensional representation of EDTA.

the electron donors to create the coordinate covalent bonds that sequester the metal ion. One-to-one stoichiometry of metal-EDTA complexes leads to four, five, or six coordination positions around the central metal ion being occupied.

Some values for the formation constants (K) of EDTA complexes involving metal cations are summarized in Table 3.1.[1] The formation reaction is given by the equation:

$$M + EDTA \rightleftarrows M \cdot EDTA$$

The larger the formation constant, K, the more likely the M·EDTA complex is to form and the fewer free metal ions remain.

Table 3.1

Formation Constants for Complexes of EDTA with Metal Cations: The Equilibrium Constants for the Complexation of Selected Metal Ions with the Sequestrant EDTA at 25°C

Cation	K	Cation	K
Ag^+	2.1×10^7	Hg^{2+}	6.3×10^{21}
Al^{3+}	1.3×10^{16}	Mg^{2+}	4.9×10^8
Ca^{2+}	5.0×10^{10}	Mn^{2+}	6.2×10^{13}
Cu^{2+}	6.3×10^{18}	Ni^{2+}	4.2×10^{18}
Fe^{2+}	2.1×10^{14}	Pb^{2+}	1.1×10^{18}
Fe^{3+}	1.3×10^{25}	Zn^{2+}	3.2×10^{16}

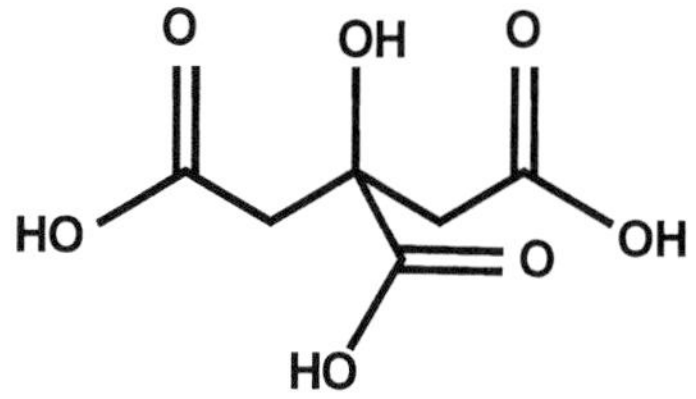

Figure 7 The chemical structure of citric acid.

A common form of EDTA used as a preservative is calcium disodium EDTA ($CaNa_2EDTA$). What metals will this form of the sequestrant scavenge effectively? The dissolution of the solid will yield calcium ions, sodium ions, and the EDTA anion. Any metal more effectively complexed than calcium will be readily scavenged, including all ions listed earlier except silver (Ag^+) and magnesium (Mg^{2+}). (In the absence of the calcium counterion, as in the case of the acid form of EDTA, chelation of calcium in the body can occur. In fact, EDTA administered orally is an FDA-approved treatment for calcium deposits in the bloodstream that lead to cardiovascular disease.)

Citric acid (Fig. 7) is another sequestrant of metal ions in foodstuffs.

KEY TERMS

chelating agent; sequestrant; coordinate covalent bond

REFERENCE

1. Skoog, D. A., West, D. M., and Holler, F. J. (1988). Formation constants for EDTA complexes. *In: Fundamentals of analytical chemistry,* 5th ed., p. 261. Saunders College Publishing, New York.

RELATED WEB SITES

"Are There Benefits behind EDTA Oral Chelation? The Reason behind Oral EDTA's Effectiveness," Dr.Gregory C. D. Young, http://vaxa.com/html/edta.html

"Chelation Therapy: The Chelation Phenomenon: A Natural Biochemical Process," Leon Chaitow, N.D., D.O., http://www.healthy.net/hwlibrarybooks/chaitow/chelther/intro/process.htm

"Chelation Therapy: The History of EDTA," Leon Chaitow, N.D., D.O., http://www.healthy.net/library/books/chaitow/chelther/intro/history.htm

"Chemical of the Week: Chelating Agents," Department of Chemistry, University of Wisconsin at Madison, http://scifun.chem.wisc.edu/chemweek/ChelatingAgents/Chelating Agents.html

WHY DOES SODA LOSE ITS CARBONATION AS ITS TEMPERATURE RISES?

The ideal serving temperature of a carbonated beverage is often stated to be near 40°F to maintain a high degree of carbonation. Many variables can affect how quickly the bubbliness of the beverage is lost, but temperature is a critical factor. The behavior of gases and liquids as a function of temperature play an important role in how refreshing we find our favorite carbonated beverage.

THE CHEMICAL ESSENCE

Although carbonated beverages represent a multibillion dollar industry in modern society, the carbonation process for producing soft drinks was first developed in the United States in the 1770s. What prompted both European and American scientists to search for viable methods to produce carbonation in beverages? The desire was attributed to the reputed healthful properties of the naturally effervescent mineral waters found at various springs throughout Europe. Today carbonation—in the form of carbon dioxide (CO_2) gas—is introduced in beverages to add taste and effervescence to appeal to both the taste buds and the eye. The added CO_2 bubbles also enhance shelf life and prevent spoilage. The carbonation process takes advantage of the particular conditions—cold temperatures and high pressures—that favor the mixing of a gas and a liquid such as water. To introduce the CO_2 gas into the water, the liquid beverage mixture is chilled below 40°F and then cascaded over a series of plates in an enclosure containing carbon dioxide gas under pressure (three to four times higher than atmospheric pressure). The amount of gas that the water will absorb during this process is enhanced by the cold temperatures and the high pressures. Whenever a bottle or can of carbonated beverage is opened, the decrease in pressure lowers the ability of the CO_2 gas to stay dissolved in the water solution. As the temperature of the beverage rises, the solubility of the dissolved CO_2 gas also decreases. The loss of effervescence is evidence of this decrease in gas solubility. This behavior is directly opposite to our experience with the solubility of solids in water. Most solids (for example, sugar or salt) dissolve more readily in hot water than in cold.

THE CHEMICAL SPECIFICS

The solubility of nearly all gases in water decreases as the temperature is increased. Furthermore, the solubility of a gas increases with the partial pressure of the gas above the surface of a liquid solution, expressed as *Henry's law:*

$$S_{gas} \text{ (moles/liter)} = k_H \times P_{gas} \text{ (atm)}$$

Both of these facts are employed in the carbonation process of sodas and beer and some sparkling wines. Low-temperature conditions and CO_2 pressures of 3 to 4 atm are used to enhance the dissolution of carbon dioxide gas in water.

The graph in Figure 8 presents the solubility of carbon dioxide in water at various temperatures and pressures.[1] The parameter used to express CO_2 solubility is the mole fraction of CO_2 in the liquid phase.

As a further note, the added tang of carbonated beverages results from the formation of the weak acid carbonic acid in water, defined by the equilibrium reaction:

$$CO_2 \text{ (g)} + H_2O \text{ (l)} \rightleftarrows H_2CO_3 \text{ (aq)} \rightleftarrows H^+ \text{ (aq)} + HCO_3^- \text{ (aq)}$$

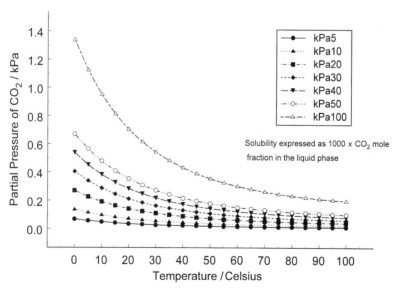

Figure 8 The solubility of gaseous carbon dioxide in water as a function of both temperature and pressure. The CO_2 solubility is expressed in terms of the mole fraction of carbon dioxide in the liquid solution.

KEYWORDS

solubility; Henry's law; partial pressure

REFERENCE

1. Lide, D.R. (Ed.). (1993). Solubility of carbon dioxide in water at various temperatures and pressures, *Handbook of chemistry and physics,* 74th ed., p. 6-7. CRC Press, Boca Raton, FL.

WHY ARE BUBBLES IN A SOFT DRINK SPHERICAL?

Imagine a stream of bubbles rapidly rising to the surface of glass of soda. What shape are these bubbles? Without any prompting, all of us would picture spherically shaped bubbles. What chemical principles preclude oval-shaped or square-shaped bubbles?

THE CHEMICAL ESSENCE

Bubbles in a soft drink primarily arise from the carbonation, or dissolved carbon dioxide gas, added prior to bottling the soda. The gas bubbles are surrounded by liquid, primarily composed of water. The water dictates the shape of the gas bubble. From an energy standpoint, water molecules prefer to interact with other water molecules and not with the carbon dioxide gas molecules. The dissimilarities in chemical structure and properties of liquid water and gaseous carbon dioxide create an unfavorable interaction of these distinct substances. To reduce the amount of interaction of water with gas, a spherical bubble forms. Why? For a given volume of space, the shape that has the lowest surface area is a sphere. By inducing the gas to form a spherical bubble, the water surrounding the bubble minimizes the surface area of the gas bubble with which it must interact. In this way, fewer water molecules must surround the gas bubble, maximizing the number of water molecules that remain in close proximity to only other water molecules.

THE CHEMICAL SPECIFICS

The spherical shape of bubbles arises from the sizable surface tension of the liquid soft drink (primarily water). The surface tension of a liquid is a measure of its resistance to increase its surface area. Liquids with relatively large intermolecular forces tend to have relatively high surface tensions, that is, high resistances to increasing their surface area. Why do the sizable intermolecular forces between water molecules necessarily dictate a spherical shape for the enclosed gas bubble? In a soft drink a water molecule at a distance from a bubble is able to experience intermolecular attractions to other water molecules surrounding it. On the other hand, a water molecule at the surface of a gas bubble has fewer neighboring liquid molecules with which it may undergo intermolecular attractions. As water molecules have strong attractive intermolecular forces (hydrogen bonding), the positioning of water molecules at the surface of a gas bubble requires some energy, because some intermolecular forces have to be overcome. For a given volume a sphere has a smaller surface area than any other shape, thus the minimum energy cost to form a bubble occurs when the surrounding liquid molecules form a spherical "enclosure" about a gas bubble.

Bubbles formed in liquids of lower surface tension than water may often be observed to have nonspherical shapes. Soap solutions, solutions of water with added surfactant that lower the surface tension of the water, can form bubbles of distorted spherical shapes.

KEY TERMS

hydrogen bonding; intermolecular forces; surface tension

WHY ARE METAL OR GLASS BOWLS PREFERRED FOR WHIPPING EGG WHITES?

Recipes for meringues often include instructions such as "In a large glass or metal mixing bowl, beat egg whites and cream of tartar until soft peaks form. Gradually beat in sugar then vanilla until stiff peaks form." Why do professional bakers and experienced cooks (as well as knowledgeable chemists!) heed the recipe's call for a metal or glass mixing bowl?

THE CHEMICAL ESSENCE

Vigorous mixing of egg whites introduces an extensive amount of air bubbles, producing a foam that retains its structure during baking. A voluminous amount of whipped egg whites is achieved by ensuring that no fat is present. Egg yolks contain fats or lipids, so the separation of yolks from whites is essential. Plastic bowls and utensils are porous and often retain fats even after washing. The surfaces of glass and metal bowls are essentially fat free, producing the most extensive degree of whipping.

THE CHEMICAL SPECIFICS

A foam is a colloidal dispersion of gas bubbles trapped in a liquid. To produce a stable foam, several characteristics of the liquid are necessary. For example, a viscous liquid facilitates the trapping of gas bubbles. The presence of a surface active agent or stabilizer that, for structural reasons, preferentially locates on the surface of the gas bubble also provides a more permanent foam. A low vapor pressure for the liquid reduces the likelihood that the liquid molecules (particularly those surrounding the bubble) will easily evaporate, thus leading to the collapse of the foam.

The albumen of egg white is a protein solution that foams readily upon whipping. Research suggests that the proteins ovomucin, ovoglobulins, and conalbumins are primarily responsible for foam formation. The proteins collect at the air-water interface of the air bubble and denature (unfold) to support the foam structure. Further denaturation through heating (baking) coagulates the proteins to result in a more stable structure. The addition of sugar during beating enhances the formation of the egg white foam due to the hygroscopic nature of sugar that retains water. (The hydroxyl groups on the sugar structure form hydrogen bonds with water.) However, sugar retards denaturation, and therefore more beating is required to reach the same extent of foam formation, particularly if sugar is added too early in the whipping process. The addition of cream of tartar, an acid (tartaric acid), lowers the pH of the protein solution, facilitating protein denaturation and coagulation. Fat, if present, would also tend to collect at the air/water interface of the bubble. However, fat, unlike protein, does not denature but coagulates. Thus, the presence of fat reduces the ability of the protein to denature and stabilize the foam.

KEY TERMS

foam; colloidal dispersion

RELATED WEB SITES

"Food Study Manual," C. Daem and D. Peabody, The School of Family & Nutritional Sciences, University of British Columbia, http://www.library.ubc.ca/ereserve/hunu201/fdmanual/

"Protein Foams—General Comments," Science of Foods, Oregon State University, http://osu.orst.edu/instruct/nfm236/foam/eggmilk.html

WHEN A FLAMBÉ DISH IS PREPARED, WHY IS THE LIQUEUR HEATED PRIOR TO LIGHTING THE FLAME?

Cherries Jubilee, Bananas Foster, Crepes Suzette, Beef Flambé, Steak Au Poivre—all are recipes that use a dramatic procedure of flaming a liqueur to introduce additional flavors into the preparation. A little knowledge of chemistry will ensure that your concoction is a delicious one!

THE CHEMICAL ESSENCE

Many recipes call for a *warmed* liqueur to be ignited with a match. This procedure, known as flambé or to flame, enhances the flavor of a dish through the process of carmelization. The ignition of the flame evaporates the alcohol and helps the flavor of the liqueur to blend into the food. Because the liqueur is to be ignited with a flame, you might wonder why the recipe instructs you to warm the liqueur prior to flaming. The liquid itself is not the substance that sustains the flame. The vapors associated with the alcohol component of the liqueur are responsible for the spectacular appearance. Heat facilitates flaming by producing vapors of the alcohol, which ignite more easily than the liquid phase.

THE CHEMICAL SPECIFICS

A fire will not occur until a *flammable* (i.e., capable of being ignited) liquid is heated above a certain temperature called the *flash point*. The flash point of a liquid is the lowest temperature at which the liquid gives off enough vapor to ignite on exposure to a flame. A cold liqueur, that is, liquid at a temperature below its flash point, does not produce enough vapors to burn. (The *vapor pressure* of a liquid increases with temperature, i.e., the amount of vapor in equilibrium with the liquid form of the substance increases as temperature is raised.) The flammable ingredient in liqueurs is ethanol with a flash point of 55°F. Cold or even room-temperature rums, brandies, and liqueurs will not support a flame for two reasons. The first reason is that the concentration of alcohol in the water is too low for the water/alcohol mixture to burn. The second reason, actually related to the first, is that the vapor pressure of the flammable alcohol over the liqueur solution is too low to be ignited. The concept of proof in alcoholic beverages relates to the amount of alcohol in the mixture. Alcoholic proof liquors less than 100 proof cannot burn. Most liqueurs are less than 100 proof. However, when a liqueur-containing mixture of less than 100 proof is heated, vapors will readily form above the surface of the skillet or pan, which will ignite if a match is brought close to the surface.

KEY TERMS

flash point; flammable; vapor pressure

WHY DO CRYSTALS FORM IN WINE OVER TIME?

A wine bottle with sediment or crystals at the bottom is not flawed. Although many consumers find such crystals to be unattractive, vintners know that formation is commonplace in wine production. What chemistry explains the origin of the crystals in wine?

THE CHEMICAL ESSENCE

The complexity of wine composition is a central reason for the vast variety of wines in the marketplace. In addition to water and ethanol, the major components, a variety of organic acids are present as well as metal ions from minerals in the skin of the grape. Initially, all of these substances remain dissolved in the bottled grape juice. As the fermentation process occurs, the increasing alcohol concentration in the wine alters the solubility of particular combinations of acid and metal ion. Unable to remain in solution, the insoluble substances settle as crystals. Because the process of red-wine making involves extended contact of the grape juice with the skins of the grapes (where the minerals are concentrated), wine crystals are more common in red wines than in white wines.

THE CHEMICAL SPECIFICS

Grapes are one of the few fruit crops that contain a significant amount of the weak organic acid known as tartaric acid, $HOOC-(CHOH)_2-COOH$. More than half of the acid content of wine is ascribed to tartaric acid. As a weak acid, tartaric acid partially ionizes in water to yield the bitartrate or hydrogen tartrate ion:

$$HOOC-(CHOH)_2-COOH \text{ (aq)} \rightleftarrows H^+ \text{ (aq)} + \\ HOOC-(CHOH)_2-COO^- \text{ (aq)}$$

The bitartrate ion can combine with potassium ion, also present in high concentrations in grapes, to form the soluble salt potassium bitartrate (also known as cream of tartar). In water sodium, bitartrate is fairly soluble: 1 gram dissolves in 162 ml of water at room temperature.[1] In alcohol solution (formed as fermentation of the wine yields ethanol), the solubility of potassium bitartrate is significantly reduced: 8820 ml of ethanol are required to dissolve 1 gram of the salt.[1] As a consequence deposits of potassium bitartrate form as the salt precipitates out of solution.

To prevent the formation of wine crystals during the bottling process, wine makers use a method known as *cold stabilization*. By lowering the temperature of the wine to 19 to 23°F for several days or weeks, the solubility of tartrate crystals is lowered, forcing the crystals to sediment. The resulting wine is then filtered off the tartrate deposit. The temperature dependence of the solubility of potassium bitartrate is readily apparent in the following comparison: although 162 ml of water at room temperature dissolves 1 gram of the salt, only 16 ml of water at 100°C are needed to solubilize the same amount of salt.[1] Recent developments employ a technique known as electrodialysis to remove tartrate, bitartrate, and potassium ions from newly fermented wine at the winery before potassium bitartrate crystals form.[2]

KEY TERMS

solubility

REFERENCES

1. Windholz, M., and Budavari, S. (Eds.). (1983). *The Merck index: An encyclopedia of chemicals, drugs, and biologicals*, 10th ed. Merck & Co., Inc., Rahway, NJ.
2. "Electrodialysis Removes Crystals but Technology Has Been ahead of Its Practicability," Steve Heimoff, *Wine Business Monthly*, May 1996, http://smartwine.com/wbm/1996/9605/bm059612.htm

RELATED WEB SITES

Thoukis, G. (1974). Chemistry of wine stabilization: A review. *In: Chemistry of winemaking*, E. & J. Gallo Winery, American Chemical Society, chap. 5, http://pubs.acs.org/books/wine.htm

Chapter 4

Connections to Space

WHY IS AN ASTRONAUT'S VISOR SO REFLECTIVE?

The bright reflection of the sun's rays on an astronaut's visor is keenly apparent in many of the photographs taken during spacewalks. The chemistry of the visor reveals the origin of its highly reflective nature.

THE CHEMICAL ESSENCE

We're familiar with mirrors—polished surfaces that divert light according to the law of reflection. In ancient times, polished castings of solid tin, bronze, copper, gold, and silver served as the first mirrors. Modern mirrors consist of a plate of glass with a thin layer of aluminum or silver deposited on the backing. Thus, glass serves as the substrate—the underlying material to which a coating is applied. The brilliant white luster of aluminum and silver metal contributes to their selection as mirror coatings. Although the decorative beauty of silver has been known for centuries, the *silvering* process of making mirrors was discovered in 1835 by the German chemist Justus von Liebig.

A typical mirror *reflects* both visible and near-infrared light. The thickness of the silver layer further delineates a mirror's function. Without the *transmission* of visible radiation, one cannot see through a silvered mirror. The coating on an astronaut's visor permits the visor to act as both a mirror and a transmitting shield. Coated with an ultrathin film of gold, infrared light reflects off the visor's surface while still permitting the astronaut to see visible light through the visor shield. The relative amount of reflection to transmittance can be controlled by the thickness of the film. A "gold-plated" visor is absolutely necessary to protect the astronaut from the infrared rays of the sun that are essentially unfiltered in space. Even the general public can benefit from this technology—gold-mirrored lenses are popular on some brands of sunglasses. Surfaces other than glass have been treated with gold for infrared protection. For example, gold tape served as a covering on the tetherlines connecting the Gemini-Titan 4 astronauts with their spacecraft during spacewalks ("extravehicular activity"). Thin films of gold also coat satellites in order to control the temperatures that could result from infrared heating in space. The exterior of the canopies of F-16 aircraft are also treated with ultrathin layers of gold metal, presumably to reduce the radar signature of the aircraft (although the purpose of the gold treatment on the cockpit is officially classified).

THE CHEMICAL SPECIFICS

An astronaut's visor is coated with thin films of the element gold. This coating reflects up to 98% of the infrared radiation incident on the visor, protecting the astronaut from the intensity of the sun's heat. The same principle dictates why gold films (as thin as even 20 picometers!)[1] are placed on the inside face of windows in office buildings, reducing heat losses in winters and reflecting infrared radiation (and thus heat) in summer. Gold is more malleable and ductile than any other metal, enabling the creation of thin films, and its high thermal conductivity (only silver and copper have higher thermal conductivities)[2] enables efficient cooling. However, the high degree of reflectivity of gold in the infrared region (98%)[3] minimizes the absorption of radiation by the gold coating deposited on the visor or glass.

KEY TERMS

gold; infrared radiation; reflectivity; thermal conductivity; reflection; transmission; silvering

REFERENCES

1. Greenwood, N. N., Earnshaw, A. (1984). Copper, silver, and gold. In: Chemistry of the Elements, chap. 28, pp. 1364–1394. Pergamon Press, New York.

2. Lide, D. R. (Ed.). (1993). Thermal and physical properties of pure metals. *In: Handbook of Chemistry and Physics,* 74th ed., pp. 12-134–12-137. CRC Press, Boca Raton, FL.
3. "Reflective Coatings for Efficient Mirrors," JML Optical Industries, Inc., http://www.jmlopt.com/ref_intro.html

RELATED WEB SITES

"Gold Occurrences," R. James Weick, Geological Survey of Newfoundland and Labrador, http://www.geosurv.gov.nf.ca/education/occgold.html
NASA Photos of Astronaut Edward H. White II during Extravehicular Activity Performed during the Gemini-Titan 4 Space Flight, http://www.ksc.nasa.gov/mirrors/images/images/pao/GT4/10074015.htm
http://www.ksc.nasa.gov/mirrors/images/images/pao/GT4/10074016.htm
http://www.ksc.nasa.gov/mirrors/images/images/pao/GT4/10074017.htm
http://www.ksc.nasa.gov/mirrors/images/images/pao/GT4/10074018.htm

WHAT IS THE SOURCE OF THE BILLOWS OF WHITE SMOKE THAT ARE SEEN WHEN THE BOOSTER ROCKETS IGNITE UPON LIFTOFF OF THE SPACE SHUTTLE?

The thunderous roar of a rocket's engines and the tremendous white clouds of smoke that accompany liftoff evoke powerful emotions in any one viewing a space shuttle launch. The chemistry of the propellant mixture is responsible for the billowy clouds as the shuttle soars into the sky.

THE CHEMICAL ESSENCE

The solid rocket boosters of the space shuttle are appropriately named for the solid propellant loaded within them. The ignition provided by the reaction of the solid aluminum powder and ammonium perchlorate powder generates a finely divided white powder known as alumina, various gases, and an extensive amount of heat. The dispersal of the white powder in the gases streaming from the boosters creates the billowy white appearance.

THE CHEMICAL SPECIFICS

The propellant mixture in each solid rocket booster of the space shuttle contains ammonium perchlorate ("the oxidizer," 69.6% by weight), aluminum ("the fuel," 16% by weight), an iron oxide catalyst (0.4 percent by weight), a polymeric binder that holds the mixture together (12.04% by weight), and an epoxy curing agent (1.96% by weight).[1] Two of the oxidation-reduction reactions that occur are the following:

$$2 \, NH_4ClO_4(s) + 2 \, Al(s) \rightarrow$$
$$Al_2O_3(s) + 2 \, HCl(g) + 2 \, NO(g) + 3 \, H_2O(g) \qquad (1)$$

$$6 \, NH_4ClO_4(s) + 10 \, Al(s) \rightarrow$$
$$5 \, Al_2O_3(s) + 6 \, HCl(g) + 3 \, N_2(g) + 9 \, H_2O(g) \qquad (2)$$

The solid white product, alumina, Al_2O_3, is dispersed in the gaseous products, creating the billowy white exhaust plumes characteristic of shuttle launches. Notice that, in the formation of alumina, three elements—aluminum, nitrogen, and chlorine—undergo changes in oxidation number. In both reactions (1) and (2), aluminum is oxidized from the zero oxidation state to the +3 oxidation state and chlorine is reduced from the +7 to the −1 oxidation state. The extent of oxidation of nitrogen differs in the two processes. In reaction (1) nitrogen undergoes a more extensive oxidation state change from −3 to +2, while in reaction (2) nitrogen experiences a more restricted oxidation from −3 to the zero oxidation state.

KEY TERMS

oxidation-reduction; oxidation state

REFERENCES

1. "Solid Rocket Boosters," NASA Space Shuttle Reference Manual, http://www.ilc-usn. kcc.ca/conversions/propuls/shtlref/srb.htm

RELATED WEB SITES

"Chapter 6. Launch Systems and Launch Sites. Section 1: Principles of Rocket Propulsion," Army Space Reference Text, U.S. Army Space Institute, Fort Leavenworth, KS, http://www.fas.org/spp/military/docops/army/ref_text/chap6im.htm
"The Space Shuttle: Solid Rocket Boosters," Student Space Awareness—National Web Team, University of Arizona Chapter of Students for the Exploration and Development of Space, http://seds.lpl.arizona.edu/ssa/docs/Space.Shuttle/srb.shtml

Chapter 5

Connections to the Outdoors and the Environment

WHAT DO METEOROLOGISTS USE TO SEED CLOUDS?

Rainmaking. Although the term invokes images of primitive rituals to influence the rain gods, modern-day meteorologists use sound chemical principles to induce precipitation. What chemistry fundamentals are instrumental in cloud seeding?

The Chemical Essence

Clouds form as masses of warm, moist air rise by convection into cooler regions, causing the water vapor to condense into liquid water droplets or ice crystals. Condensation normally occurs on microscopic particles suspended in the air, such as sea-salt particles, clay-silicate particles from land, and smoke particles produced by combustion processes such as forest fires. These particles are known as *cloud condensation nuclei*. What causes the release of precipitation from clouds? The water droplets or ice crystals must grow to sufficient size to fall from the cloud at a speed to avoid evaporation during their fall through the air to reach the ground. Many clouds composed of liquid droplets limited to a few tens of micrometers are stable for long periods of time. The collision of cloud droplets and the aggregation process that occurs to form larger particles is known as *coalescence*. In clouds with ice crystals, these particles serve as condensation nuclei.

57

Meteorologists often attempt to modify the weather to combat extreme situations such as drought. Measures such as cloud seeding are often effective when *supercooled clouds* exist, that is, clouds composed of liquid water droplets at temperatures below the freezing point of water. Atmospheric conditions routinely result in clouds at temperatures as low as $-10°C$ or $-20°C$, well below the normal freezing point of water. If ice crystal formation occurred, as expected at the given temperature, the crystals would fall out of the clouds and melt at higher temperatures closer to earth. To induce precipitation in supercooled clouds, substances are introduced to act as condensation nuclei around which water droplets coalesce.

In 1946 Irving Langmuir and Vincent J. Schaefer, chemists working at the General Electric Research Laboratories in Schenectady, New York, discovered that dry ice pellets could induce the formation of ice crystals in a cloud composed of water droplets in a deep-freeze box. The ice crystals continued to grow in size and eventually dropped to the bottom of the container. Grains of dry ice can be dispersed from airplanes to create the same effect in the atmosphere. Other substances, notably silver iodide, mimic the structure of ice and can serve as a nucleus for ice crystal formation. These substances can be introduced into clouds via aircraft or alternatively from the ground using rockets carrying a pyrotechnic substance ingrained with silver iodide. In addition, when these iodide salts are burned in air, a smoke of tiny particles is generated that can be carried upward by air currents. It could be supposed that wilderness wildfires are often extinguished by rainfall that is induced through the seeding of clouds by smoke particulates.

Seeding with silver iodide particles has also been used to mitigate or suppress the formation of hailstones in cumulonimbus clouds. By increasing the number of hail particles and thereby decreasing their size, the destructive effects are reduced. Silver iodide has also been effective at dispersing fog at airports, inducing the formation of ice crystals and the precipitation of snow. Results have generally been inconclusive, however, for seeding programs to reduce the intensity of hurricanes.

THE CHEMICAL SPECIFICS

Some of the most effective substances found for cloud seeding are solid carbon dioxide and silver iodide. As solid CO_2 at $-77°C$ drops through a supercooled cloud layer, the water droplets in the cloud instantaneously freeze. The ice crystals continue to grow in size as water vapor condenses on the solid particles, eventually reaching the dimensions necessary to fall as precipitation.

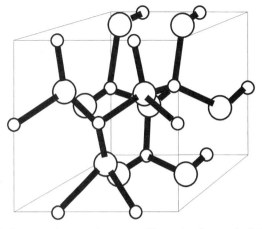

Figure 1 The ZnO or wurtzite crystal structure illustrating the tetrahedral arrangement of oxygen atoms about each zinc atom.

Why is solid silver iodide so effective at nucleating ice crystals in supercooled clouds? The answer lies in the similarity of the crystal lattice structure of silver iodide to that of ice. Although at least nine structurally different forms of ice are known, "normal" hexagonal ice predominates at the temperatures and pressures where clouds form. In this structural form of ice, each oxygen atom is surrounded by a nearly tetrahedral arrangement of four other oxygen atoms in neighboring water molecules. One of the crystal forms of silver iodide, $\beta - AgI$, has a hexagonal ZnO or wurtzite structure similar to that of ice (Fig. 1).[1] Thus, water can form ice crystals growing on the AgI seed crystals, just as if the water were growing on existing ice crystals.

KEY TERMS

supercooled liquid; crystal structure; silver iodide

REFERENCE

1. Greenwood, N. N., and Earnshaw, A. (1984). Copper, silver, and gold. *In: Chemistry of the elements,* chap. 28, p. 1376. Pergamon Press, New York.

RELATED WEB SITES

"The Discovery of Cloud Seeding," United States Bureau of Reclamation Environmental Education, http://www.usbr.gov/env_ed/snowstorm/history.htm

"Does Weather Modification Really Work?" Ric Jensen, Information Specialist, Texas Water Resources Institute, http://twri.tamu.edu/twripubs/WtrResrc/v20n2/text.html

"Irving Langmuir," Woodrow Wilson Leadership Program in Chemistry, The Woodrow Wilson National Fellowship Foundation, http://www.woodrow.org/teachers/chemistry/institutes/1992/Langmuir.html

"The Physical Basis for Seeding Clouds," Atmospherics, Inc., http://www.atmos-inc.com/weamod.html

WHY IS THE AURORA BOREALIS SO COLORFUL?

> And the Skies of night were alive with light, with a throbbing, thrilling flame; Amber and rose and violet, opal and gold it came. It swept the sky like a giant scythe, it quivered back to a wedge; Argently bright, it cleft the night with a wavy golden edge. Pennants of silver waved and streamed, lazy banners unfurled; Sudden splendors of sabres gleamed, lightning javelins were hurled. There in our awe we crouched and saw with our wild, uplifted eyes Charge and retire the hosts of fire in the battlefield of the skies.
> —Robert W. Service from "The Ballad of the Northern Lights"

THE CHEMICAL ESSENCE

Have you ever been treated to the breathtaking display of the northern lights or *aurora borealis* of the Northern Hemisphere (or the corresponding southern lights *aurora australis* of the Southern Hemisphere)? The colorful auroral rings and patches of light in the nighttime sky are a wondrous sight. Aurora was the Roman goddess of dawn, and aurora borealis and aurora australis literally mean "dawn of the north" and "dawn of the south." In fact, auroras are the final event of a chain of reactions that begin with the sun. Solar flares eject energetic particles that travel throughout interplanetary space at a high velocity (400 km/s or 250 miles/sec).[1] These particles are charged and can be trapped by the earth's magnetic field. As the particles travel to the earth's magnetic poles, they collide with oxygen (i.e., dioxygen O_2) and nitrogen (i.e., dinitrogen N_2) gases in our upper atmosphere 40 to 600 miles above Earth. These collisions result in a transfer of energy to the gaseous molecules that may even be sufficient to *ionize* the molecules (create a charged species by removing an electron, e.g., $O_2 + e^- \rightarrow O_2^+ + 2\,e^-$) or *dissociate* (separate) the molecules into individual atoms (e.g., $O_2 + e^- \rightarrow 2\,O + e^-$). The collection of energized molecules, ions, and atoms return to a less energetic state by releasing the acquired energy in the form of light. The color of the light observed depends on the identity of the species.

THE CHEMICAL SPECIFICS

An aurora is generated by streams of energetic charged particles (mostly electrons) that emanate from explosive activity, such as solar flares, on the surface of the sun. Interaction of the "solar wind" with the earth's magnetic field restricts these particles to higher latitudes. The characteristic wavelengths observed in auroras depend on the chemical identity of the energetic species created and the energy transferred from the solar particles. Highly energetic electrons can penetrate to great depths in the atmosphere, whereas low-energetic electrons only reach to the top of the thermosphere. Violet and blue light with wavelengths of 391.4 nm and 470.0 nm is emitted by N_2^+, predominantly located in the lower atmosphere. The sensitivity of the human eye generally does not readily detect the blue wavelengths. Crimson-red light at 630 nm is discharged by O_2^+. Both a greenish-yellow light at 557.7 nm (high energy) and a deep red light at 630.0 nm and 636.4 nm (low energy) are characteristic of O atoms, created via dissociation of diatomic oxygen by ultraviolet light. Yellow-green auroras arise from oxygen atoms at lower altitudes where energetic electrons penetrate. The combination of a large influx of low-energy electrons and high-altitude oxygen (200 miles up) is responsible for the rare all-red auroras.[2]

KEY TERMS

wavelength of light; energy; emission; ionization; dissociation

REFERENCES

1. "The Sun," NASA Goddard Space Flight Center, http://www-spof.gsfc.nasa.gov/Education/Isun.html
2. "The Rare Red Aurora," Article #918, Carla Helfferich, *Alaska Science Forum*, March 22, 1989, http://www.gi.alaska.edu/ScienceForum/ASF9/918.html

RELATED WEB SITES

"The Aurora Page," Department of Geological Engineering and Sciences, Michigan Technological University, Houghton, MI, http://www.geo.mtu.edu/weather/aurora/

"Auroras: Paintings in the Sky," Mish Denlinger, The Exploratorium, http://www.exploratorium.
edu/learning_studio/auroras

"The Exploration of the Earth's Magnetosphere," NASA Goddard Space Flight Center,
http://www-spof.gsfc.nasa.gov/Education/Intro.html

Photo of a Red Aurora, Alyeska Pipeline, http://www.alyeska-pipe.com/library/landscape/
aurora-red.html

"Space Weather: A Research Perspective—The Elements of Near-Earth Space," Space Studies Board, Commission on Physical Sciences, Mathematics, and Applications, National
Research Council, http://www.nas.edu/ssb/elements.html

WHY DO SEASHELLS VARY IN COLOR?

The striking beauty of seashells—their diverse colors, intricate patterns, and varied shapes—captures the interest of collectors all over the world. What role does chemistry play in the color of a seashell?

THE CHEMICAL ESSENCE

Shell collecting is a favorite pastime of many beachgoers. We're intrigued by the fascinating shapes and variety of colors of the seashells that we see around the world. But what purpose does the coloration of shells serve the inhabiting animal? Colors in seashells often serve as camouflage, provide a means of communication between species, aid in temperature regulation for intertidal species, and even serve a structural function to strengthen the shell.

The basic constituent of seashells is calcium carbonate, an insoluble compound formed from calcium ions secreted from the cells of the shellfish and carbonate ions present in seawater. But calcium carbonate is a white solid. The colors of seashells often arise from impurities and metabolic waste products that are captured in the solid shell as it is formed. Coloration is dictated by both diet and water habitat. For example, some cowries that live and feed on soft corals take on the hue of the coral species. Yellow and red colors often arise from carotenoid pigments such as β-carotene. Light refraction often generates the iridescent mother-of-pearl hues.

THE CHEMICAL SPECIFICS

The formation of a calcium carbonate shell is an example of a *precipitation reaction*:

$$Ca^{2+} (aq) + CO_3^{2-} (aq) \rightarrow CaCO_3 (s)$$

The insolubility of calcium carbonate is clearly evident from the value of the solubility product, K_{sp}, in water at 25°C: $K_{sp} = 8.7 \times 10^{-9}$. The carbonate ions are produced in seawater by the dissociation of carbonic acid that forms from the reaction of dissolved carbon dioxide gas and water:

$$H_2O (aq) + CO_2 (g) \rightarrow H_2CO_3 (aq)$$

$$H_2CO_3 (aq) \rightleftharpoons H^+ (aq) + HCO_3^- (aq)$$

$$HCO_3^- (aq) \rightarrow H^+ (aq) + CO_3^{2-} (aq)$$

Some of the pigments that may be found in seashells include melanin (brown and black hues), carotenoids (yellow and orange coloration), and pterodines (red shades). Many shellfish are capable of short-term coloration changes using pigment-containing cells known as chromatophores that are located in the deeper layers of the animal's skin. For example, the black pigment sepiomelanin (also known as sepia and the basis for sepia writing ink) isolated from the ink sac of the cuttlefish *Sepia officinalis* enables this mollusk to blend into its background.

KEY TERMS

precipitation; pigments

RELATED WEB SITES

"Use of Color and Light by Fish," Paul Maslin, Chico State University, http://www.csuchico.edu/~pmaslin/ichthy/Color.html

"Why Do Shells Have Their Colors," Gary Rosenberg, Academy of Natural Sciences, Philadelphia, http://coa.acnatsci.org/conchnet/rose0397.html

WHY ARE HYDRANGEAS PINK WHEN GROWN IN ARID REGIONS AND BLUE WHEN GROWN IN REGIONS WITH A HEAVY RAINFALL?

Hydrangeas are popular summer flowers as a consequence of their vivid color that persists throughout the season. Large blooms of pink and blue colors are common, but the predominant color varies with a number of environmental factors. What role does chemistry play in determining the hue of these magnificent flowers?

THE CHEMICAL ESSENCE

Certain varieties of hydrangeas are extremely popular for their ability to vary in color from blue to pink with growing conditions. Research has determined that the actual mechanism for color variation is determined by the concentration of aluminum compounds in the flowers.[1] In the presence of aluminum, blue flowers result; in the absence of aluminum compounds, pink flowers predominate.

Many gardening books will state that the acidity of the soil affects the color of certain varieties of hydrangeas. The soil pH affects the availability of aluminum in the soil and thereby indirectly affects the flower color. A low pH or acidic conditions will yield blue blooms; pink blossoms will be favored by a higher pH or alkaline conditions. A purple color is the result of a more moderate pH level. Potting soils with a high level of peat moss will produce blue hydrangeas. Areas with significant rainfalls also produce blue flowers, probably due to the acidic nature of the rain water. In arid regions a pink color can result from the more alkaline content of the soil.

THE CHEMICAL SPECIFICS

Soil acidity affects the availability of nutrients to the plant's root system. The solubility of a nutrient is often extremely sensitive to pH. What kinds of factors affect the pH of the soil? Some of the most important variables are the extent of rainfall, the presence of organic matter and certain microorganisms, the application of fertilizers, and the texture of the soil.

The ideal pH range to promote blue hydrangeas is 5 to 5.5, while a higher pH of 6 to 6.5 is suited for pink blooms. For plants to develop blue flowers, aluminum ion (Al^{3+}) must be free to permit complexation with a pigment molecule to produce the blue color. Low pH ensures the solubility of aluminum ion. At higher pH conditions, available aluminum ions in the soil will precipitate with hydroxide ions to form aluminum hydroxide and prevent metal complexation with the plant pigment, resulting in a pink color.

The plant pigments responsible for the variable coloration in hydrangeas is the class of pigments known as the red and blue *anthocyanins*. The anthocyanins are water soluble and display a color dependent on the acidity of their environment in the flower petal vacuoles. In acidic solution, the general structure of an anthocyanin is given by Figure 2, with a formal charge on one oxygen atom and two neighboring (i.e., *ortho*) hydroxyl groups on one of the rings.[2]

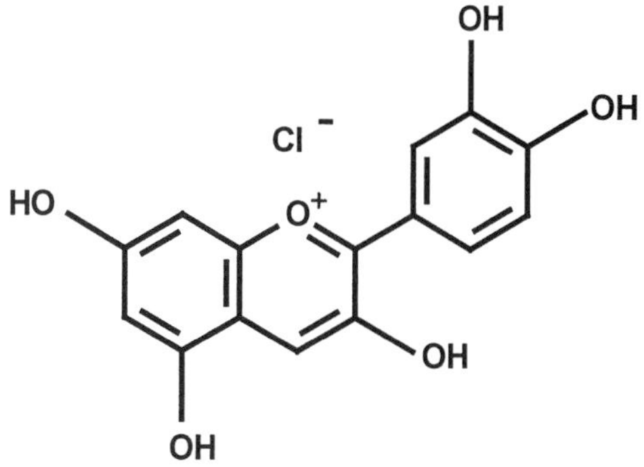

Figure 2 The general structure of an anthocyanin pigment molecule in hydrangeas.

Under these conditions, blue flowers can be achieved as a metal complex of anthocyanin (called a *metalloanthocyanin*) is stabilized through the association of a metal ion with two hydroxy groups oriented ortho to one another on the anthocyanin ring, as illustrated in Figure 3.[2,3] In basic solution, the structure of the anthocyanin (Fig. 4) no longer has *ortho*-oriented hydroxy groups. Metal complexation is no longer possible, and the flower color appears red.

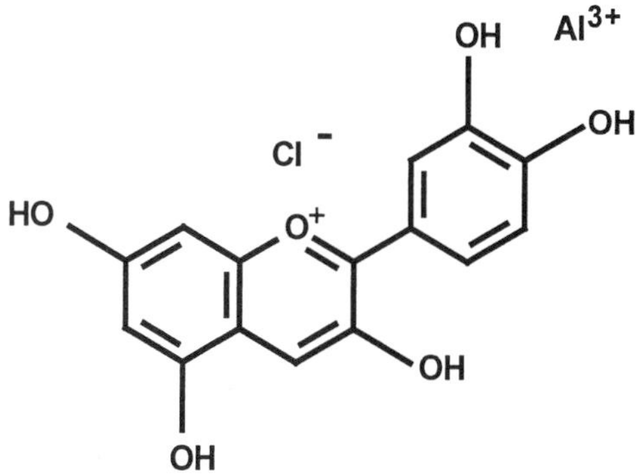

Figure 3 A metalloanthocyanin, a metal complex of anthocyanin that is stabilized through the association of a metal ion with two hydroxy groups oriented ortho to one another on the anthocyanin ring.

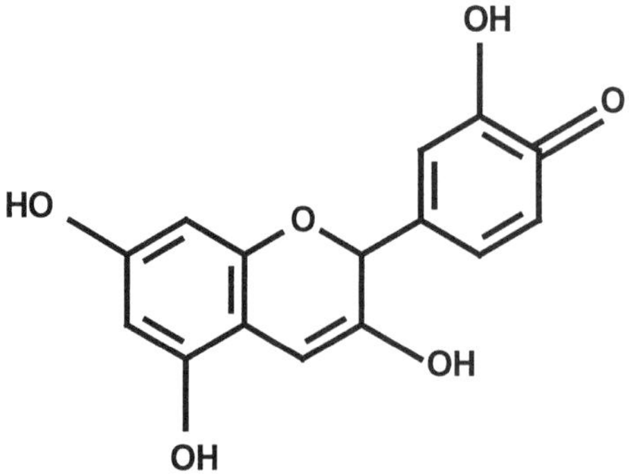

Figure 4 The structure of the anthocyanin pigment in basic solution, no longer possessing the *ortho*-oriented hydroxy groups.

KEY TERMS

acid; base; metal complexation

REFERENCES

1. "Growing Bigleaf Hydrangea," Gary L. Wade, Extension Horticulturist, The University of Georgia College of Agricultural and Environmental Sciences, Cooperative Extension Service, http://www.ces.uga.edu/Agriculture/horticulture/hydrangea.html
2. "The molecular basis of indicator color changes," F. A. Senese, Department of Chemistry, Frostburg State University, http://antoine.fsu.umd.edu/chem/senese/101/features/water2wine.shtml#anthocyanin
3. "Phenolics," Supramolecular and Physico-Chemical Laboratory, University Louis Pasteur and Chulalongkorn University, http://www.smpcl.chula.ac.th/Phenlc.htm

RELATED WEB SITES

"Hydrangea Hydrangea macrophylla," Michigan State University Extension, http://www.msue.msu.edu/msue/imp/mod10/10000171.html

"Soil pH and Fertilizers," Mississippi State Extension Service, http://ext.msstate.edu:80/pubs/is372.htm

WHY DO CITRUS GROWERS SPRAY THEIR TREES WITH WATER TO PROTECT THEM FROM A FREEZE?

During freeze warnings in certain agricultural areas, we often hear that farmers spray their crops with water to provide protection from the cold

temperatures. What chemical principles are growers applying to protect their crops?

THE CHEMICAL ESSENCE

The warm-weather climates necessary for citrus orchards generally minimize the risk of frost conditions. Nevertheless, occasionally measures are needed to protect the trees and their fruit from freezing temperatures. Citrus crops become threatened when temperatures fall below 28° for four hours or more.[1] Heating is obviously the most effective protection against frost, and often heaters are used to warm the air temperature. Microjets under trees also spray warm water onto the tree trunks to keep trees warm. As the warm water cools, heat is released to warm the air surrounding the crop. The hotter the water, the smaller the volume of water needed to provide the same degree of protection. In fact, a hot water system at temperatures near 150°F may be able to raise the orchard temperature by as much as 6°F in some conditions.[2] But the phase change of liquid water to solid ice also generates heat. Thus, spraying trees with water at conditions where freezing occurs will produce ice on the leaves and release enough heat to protect the tree and its fruit. The insulating layer of ice on plants and fruit also protects the crop by keeping the fruit temperature at 32°.

THE CHEMICAL SPECIFICS

The liquid $\rightarrow$ solid phase transition known as freezing is characterized by a negative enthalpy change (called the *enthalpy of fusion*), $\Delta H_{freezing} < 0$. At constant pressure, the enthalpy change of the freezing process is exactly equal in magnitude to the heat released as the phase transition occurs. (In other words, $\Delta H_{freezing} = q_p =$ heat transferred at constant pressure.) The *exothermic* (i.e., releasing heat) nature of the freezing process generates sufficient heat to protect the tree and its fruit.

$$H_2O \ (l) \rightarrow H_2O \ (s) \ \Delta H_{freezing} = -6.01 \ kJ/mol$$

Thus, one mole of water releases 6.01 kJ of heat as it freezes[3]; each gram of water releases 334 J upon freezing.

The use of warm water (i.e., water at temperatures above the freezing point) also generates heat as the water cools to its freezing point. The exact amount of heat released depends on the amount of water present and its temperature according to the formula:

$$\Delta H_{cooling} = C_{p,water(l)} \ \Delta T = C_{p,water(l)} \ (T_{liquid \ water} - T_{298 \ K})$$

The heat capacity of liquid water at 20°C, for example, is 4.2 J g^{-1} K^{-1}.[4] Thus, for every degree above the freezing point of water, one gram of water releases 4.2 J upon cooling one degree.

KEY TERMS

phase transition; heat capacity; enthalpy of fusion

REFERENCES

1. "Irrigation as a Tool for Frost and Freeze Protection of Agricultural Crops," Timothy W. Appelboom, Department of Biological and Agricultural Engineering, North Carolina State University, http://www2.bae.ncsu.edu/courses/bae572/SpecialReports/appelboom/paper1.html
2. "Hot Water Could Protect Trees from Cold Damage," Geraldine Warner, *Good Fruit Grower,* February 15, 1997, http://www.goodfruit.com/archive/Feb15-97/special4.html
3. Lide, D. R. (Ed.). (1993). Enthalpy of fusion. *In: Handbook of Chemistry and Physics,* 74th ed., pp. 6-116—6-125. CRC Press, Boca Raton, FL.
4. Lide, D. R. (Ed.). (1993). Properties of water in the range 0–100°C. *In: Handbook of Chemistry and Physics,* 74th ed., p. 6-10. CRC Press, Boca Raton, FL.

RELATED WEB SITES

"Orchard Smudge Pots Cooked Up Pall of Smog," South Coast Air Quality Management District, http://www.aqmd.gov/monthly/smudge.html

WHY DOES A MIXTURE OF HYDROGEN PEROXIDE AND SODIUM BICARBONATE DEODORIZE A DOG THAT HAS BEEN SPRAYED BY A SKUNK?

One pet owner recently prescribed the following home remedy to eliminate skunk odor: Add 1/4 cup of baking soda and 1 teaspoon of liquid detergent to 1 quart of hydrogen peroxide. Soak a rag with the solution and saturate the affected areas, rubbing it in.[1] *What valuable chemistry is operating to deodorize your dog?*

THE CHEMICAL ESSENCE

The odor from skunk spray originates from a substance that belongs to a family of compounds with many odiferous members. This class of substances, known as *thiols* or *mercaptans,* is responsible for the delectable aroma of freshly brewed coffee (an ingredient called *furfurylthiol*), the repugnant smell of rotten eggs (a substance known as *hydrogen sulfide*), and the warning odor of natural gas for leak detection (an odorant called *2-methyl-2-propanethiol*). The sensitivity of the human nose to these compounds, often at levels as low as 20 parts per billion, contributes to their reputation as pungent odors. By a chemical conversion of a thiol to another class of compound, the disagreeable skunk scent can be removed.

THE CHEMICAL SPECIFICS

Skunks deter predators by release of a liquid spray containing seven major volatile components[2] classified as *thiols* (compounds containing the $-SH$ functional group) and acetate derivatives of thiols (characterized by the $-SC(O)CH_3$ functionality). In particular, two of the more odiferous components responsible for the strongly repellent odor of the skunk's secretion are 2-butene-1-thiol (Fig. 5) and 3-methylbutane-1-thiol (Fig 6).[2]

To deodorize a pet sprayed by a skunk, the noxious compound must be converted to an odorless one. Using an alkaline solution of 3% hydrogen peroxide and sodium bicarbonate, the thiol compound (denoted R-SH) can be oxidized to a disulfide compound.

$$2 \text{ RSH} \rightarrow \text{RS-SR} + 2 \text{ H}^+ + 2 \text{ e}^-$$

or in the presence of an alkaline substance such as detergent:

$$2 \text{ RSH} + 2 \text{ OH}^- \rightarrow \text{RS-SR} + 2 \text{ H}_2\text{O} + 2 \text{ e}^-$$

Here hydrogen peroxide acts as the *oxidizing agent:*

$$\text{H}_2\text{O}_2 \text{ (aq)} + 2 \text{ e}^- \rightarrow 2 \text{ OH}^- \text{ (aq)}$$

Thus, the overall reaction is written as follows:

$$2 \text{ RSH} + \text{H}_2\text{O}_2 \text{ (aq)} \rightarrow \text{RS-SR} + 2 \text{ H}_2\text{O}$$
odoriferous   odorless

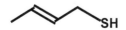

Figure 5   2-butene-1-thiol, an odiferous component of a skunk's secretion.

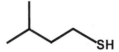

Figure 6   3-methylbutane-1-thiol, an odiferous component of a skunk's secretion.

KEY TERMS

thiol; oxidation-reduction

REFERENCES

1. "Skunk Odor Removal," http://www.hardlink.com/~gbuzzell/skunk.htm
2. "Chemistry of Skunk Spray," Professor William F. Wood, Department of Chemistry, Humboldt State Univeristy, http://sorrel.humboldt.edu/~wfw2/chemofskunkspray.html

RELATED WEB SITES

"Chemical Composition of Striped, Spotted, and Hog-Nosed Skunk Musk," James Firmiss, University of Wisconsin, Madison, http://granicus.if.org/~firmiss/m-d/skunk-chem.html
"Deodorize Skunk Spray: Neutralize skunk odor," Professor William F. Wood, Department of Chemistry, Humboldt State University, Arcata, CA, http://www.humboldt.edu/~wfw2/deodorize.shtml
"Skunk and Opossum Page," James Firmiss, University of Wisconsin, Madison, http://granicus.if.org/~firmiss/m-d.html

IF NATURAL GAS IS ODORLESS, WHY CAN WE DETECT THE ODOR OF A GAS LEAK?

Clean, reliable, and energy efficient; odorless, colorless, combustible— these attributes accurately characterize the form of energy known as natural gas. Nevertheless, we're constantly advised to contact the gas service company whenever we detect the smell of gas as a precautionary measure to locate leaks. An examination of the chemical composition of natural gas can explain this paradox.

THE CHEMICAL ESSENCE

Natural gas is actually a mixture of several combustible gases found in deposits in the earth's crust. Natural gas is comprised of a mixture of low molecular weight *hydrocarbons* (compounds comprised of the two elements carbon and hydrogen), primarily methane (CH_4) and ethane (C_2H_6). These gases are indeed odorless and colorless. However, other more pungent gaseous constituents may be present in natural gas. Some of these gases are inherently present in gas reservoirs as a consequence of the decay of the organic matter in the fossils of plants and animals that produced natural gas millions of years ago. One such gas, hydrogen sulfide—a member of the *mercaptan* or *thiol* family—has the odor of rotten eggs. Other odorous gases are deliberately added to natural gas by your gas service company as a safety precaution to facilitate the detection of both indoor and outdoor gas leaks.

THE CHEMICAL SPECIFICS

The hydrocarbon mixture in natural gas reservoirs primarily contains the low molecular-weight gaseous compounds methane (CH_4) and ethane (C_2H_6), both of which are gaseous under atmospheric conditions. Heavier hydrocarbons such as propane (C_3H_8), butane (C_4H_{10}), pentane (C_5H_{12}), and hexane (C_6H_{14}) are also likely.[1] Below the earth's surface, the high pressures ensure that even pentane and hexane (normally gases at atmospheric pressure) exist as liquid species known as liquefied petroleum gas. As these substances rise to the earth's surface, the lower pressures vaporize them to produce petroleum gas.

Other gases are often commonly present in natural gas deposits, including hydrogen, nitrogen, and carbon dioxide. Volcanic activity and the decay of organic matter also produce trace amounts of hydrogen sulfide. Even the noble gases argon and helium are constituents of natural gas, produced from natural radioactive disintegration of radioisotopes of potassium (Ar) and thorium and uranium (He). Of these component gases, the extremely penetrating odor of even a small amount of naturally occurring hydrogen sulfide, H_2S, imparts a detectable odor to natural gas. The human nose is quite sensitive to sulfur-containing compounds in the *thiol* class (containing the −SH functional group); activation of about only 40 receptor cells by 9 thiol molecules each is sufficient to produce an odor sensation.[2] As noted, odorants are also added to natural gas to enhance the detection of leaks. The compounds ethanethiol (Fig. 7), CH_3CH_2SH, and 2-methyl-2-propane-

Figure 7   Ethanethiol, CH_3CH_2SH, an example of an odorant and warning agent for natural gas.

Figure 8   2-methyl-2-propanethiol or *tert*-butyl mercaptan, an example of an odorant and warning agent for natural gas.

thiol or *tert*-butyl mercaptan (Fig. 8) are examples of odorants and warning agents for natural gas. These substances are liquids at atmospheric pressure (*normal boiling points,* i.e., boiling points at 1 atm, of 35 and 64°C, respectively, for ethanethiol and 2-methyl-2-propanethiol[3]) but can exist as gases under pressure at room temperature.

KEY TERMS

hydrocarbon; thiol

REFERENCES

1. "Natural Gas, Manufactured Gas & Liquefied Gas Analysis: Part I—Background," Atlantic Analytical Laboratory, Inc., http://www.test-lab.com/gasone.htm
2. "Sensory Reception: Animal Sensory Reception: Chemoreception: Chemoreception in the Vertebrates: General Vertebrate Chemoreception: Process of Olfaction." *Encyclopedia Britannica Online,* http://www.eb.com:180/cgi-bin/g?DocF=macro/5005/71/47.html
3. "ChemFinder," Cambridge Soft Corp., http://chemfinder.camsoft.com/

RELATED WEB SITES

"Chemical Compounds: Organic Compounds: Organic Sulfur Compounds: Organic Compounds of Bivalent Sulfur," *Encyclopedia Britannica Online,* http://www.eb.com:180/cgi-bin/g?DocF=macro/5009/2/210.html

"Facts about Natural Gas," American Petroleum Institute Education, http://www.api.org/edu/factsgas/factsgas.htm

WHAT DOES pH STAND FOR?

In countless situations, the relative acidity or alkalinity of a substance or system is of critical importance. Agricultural conditions, lung and kidney

function, water quality, food preservation—all are circumstances in which quantitative measures of acidity or basicity enable proper maintenance and regulation of vital processes. The term pH *is widely used as an expression of acid/base content, but what is the origin of this nomenclature?*

THE CHEMICAL ESSENCE

The pH of a substance, usually a solution, is a quantitative measure of the acidity of the substance. The Swedish chemist Svante Arrhenius proposed in the late 1880s that acids were substances that delivered hydrogen ions to a solution. In 1904 H. Friedenthal recommended that acidity be expressed in terms of the concentration of the hydrogen ion present. In 1909 the Danish biochemist Soren P. L. Sorenson published his classic paper, "Enzyme Studies II. The Measurement and Meaning of Hydrogen Ion Concentration in Enzymatic Processes," in the journal *Biochemische Zeitschrift.*[1] In this work he introduced a quantitative expression for the acidity of a solution. Interestingly, his designation for the parameter was written P_H, not *pH*. Mathematically speaking, Sorenson defined pH as the negative of the logarithm (base 10) of the hydrogen ion concentration expressed as a power of 10. Operationally, Sorenson showed that the pH of a solution can be easily expressed in a manner that reveals the relative acidity of the system through the *power of the hydrogen* ion concentration (or pH):

> The value of the hydrogen ion concentration will accordingly be expressed by the hydrogen ion. . . , and the factor will have the form of a *negative power of 10. . . .* I will explain here that *I use the name "hydrogen ion exponent" and the designation P_H for the numerical value of the exponents of this power.*[1]

THE CHEMICAL SPECIFICS

Sorenson chose to write the concentration of hydrogen ions as a power of 10. For example, an aqueous solution that contained 5.00×10^{-2} M hydrogen ions (H^+) would be denoted as containing the mathematically equivalent value of $10^{-1.30}$ M H^+ ions. Sorenson designated the "P_H" or pH of the solution as the numerical value of the negative exponent of 10. Thus, a pH of 1.30 would be ascribed to the solution. In other words,

$$[H^+] = 10^{-pH}$$

The concept of using the base 10 logarithm to express the magnitude is a widespread practice today. Equilibrium constants of chemical reactions are often noted or compared as pK values where $pK = -\log_{10}$ (magnitude of equilibrium constant). For example, the extent of dissociation of acetic acid, the acid in vinegar, is quantified by an equilibrium constant of 1.8×10^{-5}. Here, then, $pK = -\log_{10} (1.8 \times 10^{-5}) = 4.74$.

KEY TERMS

pH; logarithm; acidity; concentration

REFERENCES

1. "Classic Papers in Chemistry," S. P. L. Sorenson, "Enzyme Studies II. The Measurement and Meaning of Hydrogen Ion Concentration in Enzymatic Processes," *Biochemische Zeitschrift* 21, 131–200 (1909), http://dbhs.wvusd.k12.ca.us/Chem-History/Sorenson-article.html

RELATED WEB SITES

"pH Values of Foods," Jonathan Stott, http://www.stott.demon.co.uk/othersci/phfood.htm
"pH Values of Various Foods," U.S. Food & Drug Administration, Center for Food Safety & Applied Nutrition, http://vm.cfsan.fda.gov/~mow/app3a.html
"Soren Sorenson and the pH Scale," John L. Park, http://dbhs.wvusd.k12.ca.us/AcidBase/pH.html

Chapter 6

Connections to the Office

HOW DO SCRATCH-AND-SNIFF ADS WORK?
HOW DOES CARBONLESS COPY-PAPER WORK?

The clever combination of chemistry and microsphere technology enables advertisers to take advantage of the incredible marketing power of scent. These innovations have also modernized paper-based record keeping and business transactions.

THE CHEMICAL ESSENCE

These two seemingly dissimilar applications have a common basis—both are examples of the *pressure-sensitive* release of a chemical. How are these products designed? Tiny spherical capsules (*microcapsules* or *microspheres*) with a glass or polymer shell are filled with a liquid core and glued onto paper. For a scratch-and-sniff ad, the core of the microcapsules contains a liquid with the desired scent; for carbonless paper, a liquid ink or dye is encapsulated within the microspheres that adhere to the underside of the paper. When the paper ad is scratched, the shell of the microsphere is broken. By exposing the liquid cavity, the scent of the enclosed perfume or fragrance is easily detected. Similarly, when pressure is applied to the copy paper through a pen or typewriter, the encapsulated colorless dye is released and reacts with an acidic chemical on the paper underneath to darken and transfer ink to the underlying paper.

Microencapsulation is the term used for the shielding of solid, liquid, or gaseous active ingredients by enclosure of the active component within a protective shell. To release the contents of the microcapsule, four techniques are often employed. Scratch-and-sniff cards and carbonless copy paper rely on the *mechanical rupture* of the microsphere shell induced by pressure. Timed-release medicines and agrochemicals are often based on *water-soluble polymeric coatings* that dissolve over time. (See Chapter 1, "How does a timed-release medicine work?") Sustained release of medications is also achieved via *diffusion* of the active components through microcapsule walls. *Thermal breakdown or melting* of a solid microcapsule wall by increased temperature is also a release mechanism, used to liberate fat-encapsulated baking soda in packaged baking mixes, for example.

The Chemical Specifics

The polymer and glass microspheres employed in the pressure-sensitive release of chemicals range in size from 1 μm to 1 mm in diameter. (For comparison, a human hair is typically 80 to 100 μm in diameter.)[1] A scanning electron micrograph illustrating the morphology of the particles appears in Figure 1.[2]

A variety of techniques are employed to achieve microencapsulation. One of the earliest methods, developed in the 1930s, is known as *coacervation* or *phase separation*. This method reproducibly applies a uniform, thin, polymeric coating to small particles of solids or to droplets of pure liquids and solutions. To apply this method, four components are necessary: the material to be coated (core), a wall-forming polymer, a suitable solvent (liquid manufacturing vehicle), and a coacervation (phase-separation) inducer.[3] For water-soluble core materials, a polymer soluble in a nonpolar solvent such as cyclohexane must be chosen (e.g, ethylcellulose). For core substances not miscible in water, a water-soluble polymer (e.g., gelatin) is required. The polymer must also be chemically compatible with the core material and be capable of forming a cohesive film on the core surface. Selection of an appropriate polymer, whether natural or synthetic, should also address the needs for a coating material with the requisite strength, flexibility, impermeability, and stability.

In the coacervation process, the core substance is first added to a homogeneous solution of the selected solvent and polymer. Mechanical agitation is used to disperse the immiscible core to create tiny droplets suspended in solution (i.e., an emulsion). The coacervation or phase separation phe-

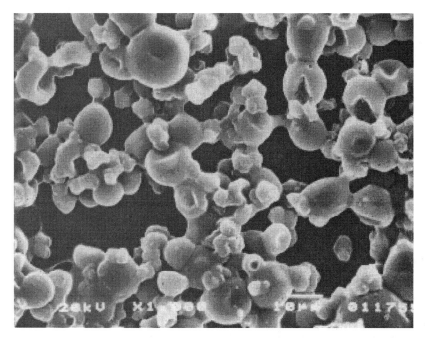

Figure 1 A scanning electron micrograph, magnified 1000 times, illustrating the morphology of the particles in a pressure-sensitive scratch-and-sniff ad. Courtesy of Museum of Science, Boston.

nomenon is then induced by several means, such as changing the temperature and/or acidity of the polymer solution or adding salts, nonsolvents, or incompatible (immiscible) polymers to the polymer solution. For example, by adjusting the acidity of the system through the addition of an acid, a phase separation may be induced. In other words, two immiscible liquid phases are created with different amounts of the solubilized polymer in each phase. The supernatant phase has low polymer concentrations, whereas the coacervate phase has a relatively high concentration of the polymers. The polymer is selected so that the coacervate phase preferentially adsorbs onto the surface of the dispersed droplets to form the shell of the microcapsules. Once the fluid film of polymer is deposited from the coacervate onto the core, the shell must be solidified. Cooling and further chemical reaction with a cross-linking agent such as formaldehyde hardens the microcapsule walls. The microcapsules are then separated by settling or filtration, and then they are washed, filtered and dried.

KEY TERMS

microencapsulation; microsphere

REFERENCES

1. "Which Object Is Closest in Length to a Micrometer?" from "The Advanced Light Source: A Tool for Solving the Mysteries of Materials," Lawrence Berkeley National Laboratory, http://www.lbl.gov/MicroWorlds/ALSTool/micrometer.html
2. "Scratch-and-Sniff Paper," The Museum of Science, Boston, http://www.mos.org/sln/sem/scratch.html
3. "Taste-Masking of Oral Formulations," Eurand Research and Development, http://www.eurand.com/taste.html

RELATED WEB SITES

"The Art and Science of Microencapsulation," J. Franjione and N. Vasishtha, *Technology Today,* http://www.swri.org/3pubs/ttoday/summer/microeng.htm

"The Carbonless Copy Paper Project," University of Florida, Department of Environmental Engineering, http://www.enveng.ufl.edu/homepp/schmidt/carbonlesscopy.htm

"Microencapsulation," Faculty of Pharmaceutical Sciences, Chulalongkorn University, http://www.pharm.chula.ac.th/microcap/MIC.HTM

"Preparation of Polymer Microcapsules with Liquid Cores," Andrew Loxley, Department of Chemistry, University of Bristol, http://www.tlchm.bris.ac.uk/vincent/al/al.htm

"Review of Pharmaceutical Controlled Release Methods and Devices," Paul A. Steward, http://www.initium.demon.co.uk/rel_nf.htm#Drug_Deliv

HOW DO "REPOSITIONAL SELF-ADHESIVE PAPER NOTEPAD SHEETS" (e.g., Post-It Notes) WORK?

THE CHEMICAL ESSENCE

Office supplies and office communication have been revolutionized by the introduction of the 3M product Post-It Notes in 1980. The temporary adhesive of these self-sticking notes actually was initially rejected as a glue by its inventor, 3M chemist Spencer Silver, because of its impermanence. But 3M researcher Art Fry found a niche for the adhesive—as an adhesive for a temporary bookmark for his choir hymnal.[1] Key to the performance of these removable, repositional adhesive products is the application of the adhesive to the backing of a note sheet via tiny microspheres rather than a continuous film. With an average particle diameter of 25 to 45 microns,

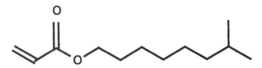

Figure 2 The chemical structure of isooctyl acrylate, a monomer in a free-radical polymerization reaction.

the microsphere adhesives form a discontinuous layer that assists in retaining the ability to reapply the note to new surfaces. Traditional adhesive tape contains particles of smaller dimension (typically 0.1 to 2.0 microns) that coalesce to form a continuous film that limits removal.[2]

THE CHEMICAL SPECIFICS

The tacky polymeric microspheres that comprise the pressure-sensitive adhesive layers of repositionable notes are patented inventions. One such material (U.S. Patent 5 714 237)[3] is prepared by a *free-radical polymerization reaction* of isooctyl acrylate (Fig. 2) in the presence of polyacrylic acid with a *chain-transfer agent* (dodecanethiol) (Fig. 3), an *initiator* (monomer soluble bis-(t-butylcyclohexyl)peroxycarbonate (Fig. 4), and a detergent (ammonium lauryl sulfate) (Fig. 5). The adhesive polymeric composition is recovered and blended with a solids coating mixture to create the microspheres.

Free-radical polymerization reactions are also known as chain-growth or addition polymerization reactions. Let's look a chain-growth polymerization reactions in more detail. As the name suggests, these processes convert small monomer molecules to larger polymers by the successive addition of monomer molecules onto the reactive ends of a growing polymer. An initiator is required in chain-growth polymerization reactions. The function of the initiator is to react with the monomer to form another reactive compound, thereby beginning ("initiating") the linking process. Under the

Figure 3 The chemical structure of dodecanethiol, a chain-transfer agent in a free-radical polymerization reaction.

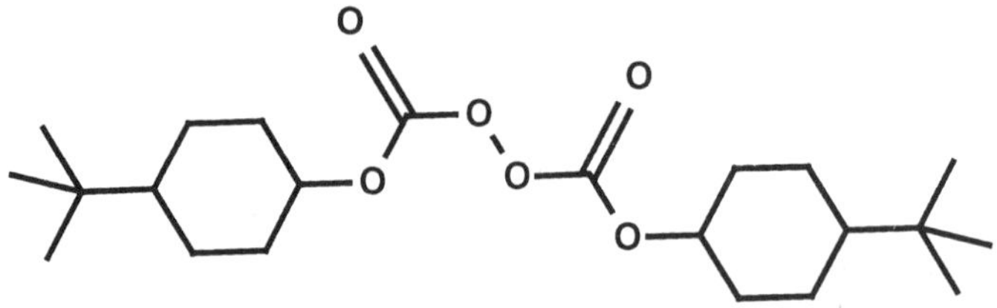

Figure 4 The chemical structure of di-(4-tert-butylcyclohexyl)peroxydicarbonate, an initiator in a free-radical polymerization reaction.

proper experimental conditions, the peroxycarbonate initiator easily cleaves at its oxygen-oxygen bond to create unstable species called free radicals (or simply radical) species that are characterized by an unpaired electron:

$$C_6H_{11}OC(O)OOC(CH_3)_3 \rightarrow C_6H_{11}OC(O)O\cdot + \cdot OC(CH_3)_3$$

One of these free radicals may decompose further by releasing carbon dioxide and generating a new free radical:

$$C_6H_{11}OC(O)O\cdot \rightarrow C_6H_{11}O\cdot + CO_2 \text{ (g)}$$

These reactive species initiate the linking process by adding to the double bond of the chosen acrylate monomer (isooctyl acrylate) to create another reactive radical species. This linking process may be represented by the following:

$$I\cdot + H_2C = CHX \rightarrow \cdot H_2C\text{-}CHI\cdot$$

This new free radical adds to the double bond of another monomer molecule, growing the polymer chain. The polymerization process ends as the unpaired electrons of two free radicals combine to form a single bond:

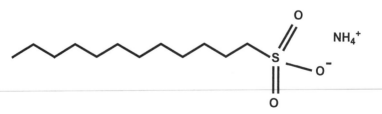

Figure 5 The chemical structure of ammonium lauryl sulfate, a detergent.

$$R - CH_2 - CHX \cdot + \cdot CHX - CH_2 - R \rightarrow$$
$$R - CH_2 - CHX - CHX - CH_2 - R$$

KEY TERMS

monomer; polymer; initiator; free radical; free-radical (chain- growth, addition) polymerization

REFERENCES

1. "Post-it Notes Become a Best-Selling Office Product," 3M, http://www.mmm.com/profile/backgrnd/postit.html
2. "StikWithit: Repositional Adhesive: Microsphere Technology," Advance Polymers Technology, http://www.padmaker.com/stik-withit.htm
3. Patent watch (1998, April). *Chemtech*, 25.

RELATED WEB SITES

"Free Radical Vinyl Polymerization," Department of Polymer Science, University of Southern Mississippi, http://monomer.psrc.usm.edu/macrog/radical.html

"Fundamental of Polymer for Fiber Applications," American Fiber Manufacturers Association, http://www.fibersource.com/f-tutor/poly.htm

"Polymer Synthesis, Polymers, and Liquid Crystals," Virtual Textbook, Case Western University, http://plc.cwru.edu/tutorial/enhanced/files/POLYMERS/synth/synth.htm

"3M Chemist Wins Creative Invention Award," http://www.mmm.com/front/silver/index.html

HOW DO CORRECTION FLUIDS LIKE LIQUID PAPER AND WHITE-OUT WORK?

Typewriter correction fluid was first invented by Bette Nesmith Graham.[1,2] *Her early version for the product consisted of a white tempura paint to correct mistakes. Looking for a mixture that was both quick-drying and barely detectable, she discovered a formulation that she would later sell to the Gillette Company in 1979. The formula was modified slightly in the 1980s. What is the chemistry of Liquid Paper?*

THE CHEMICAL ESSENCE

Paper correction fluids contain two key ingredients—a white (or colored) pigment and a volatile fluid solvent. The pigment is initially dissolved in the solvent, but, upon application to a surface, the solvent readily evaporates. A solid residue of pigment remains. If solvent evaporates from an open bottle of correction fluid, additional solvent in the form of correction fluid thinner can be added to redissolve the solid pigment.

THE CHEMICAL SPECIFICS

White-colored correction fluids like Liquid Paper and White-Out contain the pigment titanium dioxide (TiO_2) and the volatile solvent 1,1,1-trichloroethane or methyl chloroform (CCl_3CH_3).[3] The volatile substances in the correction fluid contribute 50% of the total volume of the product.[3] Correction fluid thinner is simply 100% 1,1,1-trichloroethane solvent[4], added to redissolve any solidified titanium dioxide.

KEY TERMS

volatile; solvent

REFERENCES

1. "Bette Nesmith Graham (1924–1980): Liquid Paper," Inventor of the Week Archives, The Lemelson-MIT Prize Program, http://web.mit.edu/invent/www/inventorsA-H/nesmith.html
2. "Ordinary People, Extraordinary Ideas," http://www.aas-world.org/Youth_Sciences/NHYIP/Meant_to_Invent_/Teacher_Guide/Ordinary_People/body_ordinary_people.html
3. "Liquid Paper Correction Fluid, White; Material Safety Data Sheet," http://www.biosci.ohio-state.edu/~jsmith/MSDS/LIQUID%20PAPER%20CORRECTION%20FLUID%20WHITE.htm
4. "Liquid Paper Correction Fluid Thinner, MSDS Sheet," http://www.biosci.ohio-state.edu/~jsmith/MSDS/LIQUID%20PAPER%20CORRECTION%20FLUID%20THINNER.htm

WHY DOES DISAPPEARING INK DISAPPEAR?

Chances are, you've been fascinated by the gradual fading of color when a message written with seemingly standard ink dissipates right before your eyes. Why should a chemist not be fooled by the disappearing ink or invisible ink in a magician's bag of tricks?

THE CHEMICAL ESSENCE

Whether for a class demonstration, a practical joke, or perhaps a clandestine activity, disappearing ink is a fascinating substance. What is the secret to its action? One formulation of disappearing ink contains a common acid-base indicator, that is, a substance that by its color shows the acidic or basic nature of a solution. One acid-base indicator that shifts from a colorless hue under acidic conditions to a deep blue color in alkaline solutions is thymolphthalein. If the indicator starts off in a basic solution, perhaps containing sodium hydroxide, the typical blue color of an ink is perceived. How does the ink color disappear? This behavior is dependent on the contact of the ink with air. Over time, carbon dioxide in the air combines with the sodium hydroxide in the ink solution to form a less basic substance, sodium carbonate. The carbon dioxide also combines with water in the ink to form carbonic acid. The indicator solution responds to the production of acid and returns to its colorless acidic form. A white residue (sodium carbonate) remains as the ink dries.

THE CHEMICAL SPECIFICS

Thymolphthalein (Fig. 6) (also known as 2′,2″-Dimethyl-5,5-di-iso-propylphenolphthalein) is an acid-base indicator that is colorless in its acidic

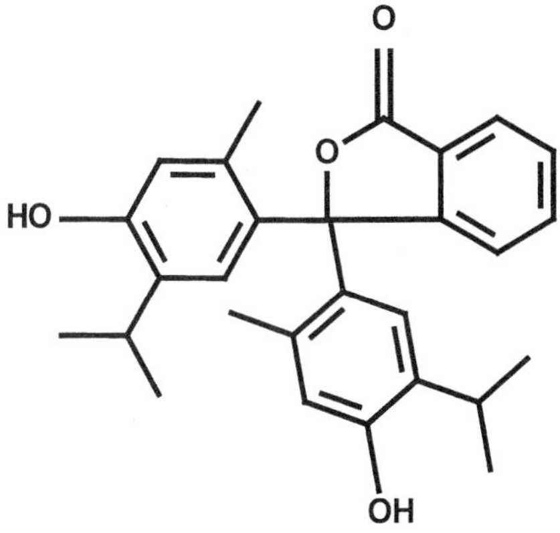

Figure 6 The chemical structure of thymolphthalein (2′,2″-Dimethyl-5,5-di-iso-propylphenolphthalein) is an acid-base indicator that is colorless in its acidic form and deep blue in its basic form.

form and deep blue in its basic form. The equilibrium between the acidic and basic forms of the indicator may be represented as follows:

$$HIn \text{ (colorless)} \rightleftarrows H^+ + In^- \text{ (deep blue)}$$

The acidic form of the indicator (HIn) retains the hydrogen on each hydroxyl group; the conjugate base form of the indicator (In^-) contains one ionized hydroxyl group (O^-). The pK_a value for the acid ionization is 9.9; thus, $K_a = 10^{-9.9} = 1.3 \times 10^{-10}$.[1] A discernible color change is noted when the pH of an aqueous solution of the indicator is in the range of 9.4 to 10.6.

Disappearing ink can be prepared by first dissolving solid thymolphthalein in ethanol, adding water, and then adjusting the pH with sodium hydroxide solution.[2] The deep blue color of the basic form of the indicator is readily apparent. Applying the ink to paper increases its exposure to carbon dioxide in air. Two chemical reactions occur. Carbon dioxide and sodium hydroxide react to form the salt sodium carbonate:

$$2 NaOH \text{ (aq)} + CO_2 \text{ (aq)} \rightarrow Na_2CO_3 \text{ (s)} + H_2O \text{ (l)}$$

Carbon dioxide and water also combine to form carbonic acid:

$$CO_2 \text{ (aq)} + H_2O \text{ (l)} \rightleftarrows H_2CO_3 \text{ (aq)}$$

The partial ionization of carbonic acid produces hydronium ion, H^+, driving the indicator equilibrium to the weak acid form. A colorless solution results. As the water in the ink evaporates, the white residue of sodium carbonate remains.

KEY TERMS

acid-base indicator; conjugate base; ionization

REFERENCES

1. "Properties of Aqueous Acid-Base Indicators at 25°C," *Chemical Sciences Data Tables,* James A. Plambeck, 1996, http://www.compusmart.ab.ca/plambeck/che/data/p00433.htm
2. "Disappearing Ink," A Chemical Demonstration by Scott Hertting, http://nhs.njsd.org/users/s/sherttin/ink.htm

HOW DOES AN AUTOMATIC FIRE SPRINKLER HEAD OPERATE?

Many office buildings rely on ceiling sprinkler systems as effective means to extinguish a fire. How are these systems activated? The heat generated by a fire is the key to the operation of these water sprinklers.

THE CHEMICAL ESSENCE

Automatic fire sprinklers are inserted into a network of piping that contains water under pressure. The sprinklers operate individually by heat activation. As a fire's intensity increases, the heat of a fire builds, exposing the sprinkler head to elevated temperatures generally between 135 to 225°F. A typical operating temperature for a sprinkler head is 165°F,[1] although temperature ratings as low as 135°F and as high as 360°F are available.[2] When the operating temperature is reached, either a solder link melts or a liquid-filled glass bulb shatters. In the solder link operating mechanism, two pieces of metal are held together with a special solder, creating the "link." The link is connected to a mechanical mechanism that holds a cap or plug in place in the water pipe. As solder melts at a specific temperature, the cap over the orifice or nozzle is ejected by water pressure. When the sprinkler head is opened, water is directed through a deflector to create a spray that is released directly over the source of heat.

In glass bulb sprinklers, liquid within a fragile glass bulb expands in volume as its temperature is raised by the heat of the fire. The glass bulb also holds a cap or plug in place in the water pipe. At a particular temperature, the volume has increased enough to shatter the glass bulb, triggering the discharge of water. Also used are chemical pellets that melt at a predetermined temperature.[3]

Fusible alloys are not only used in fire sprinkler heads but also in thermal alarms and in fuses interrupting an electrical circuit when the current becomes excessive.

THE CHEMICAL SPECIFICS

Alloys are mixtures of metals combined to obtain specific characteristics and enhanced properties for a particular application. The term *fusible metals* or *fusible alloys* denotes a group of alloys that have melting points below that of tin (232°C, 449°F). Most of these substances are mixtures of metals that by themselves have relatively low melting points, such as tin, bismuth (m.p. 275°C), indium (157°C), cadmium (321°C), and lead (327°C). Table 6.1 illustrates the dependence of melting point on the composition of several fusible alloys.[4] Note that all combinations of metals yield an alloy with a significantly lower melting point than any of the individual metals. Even more extensive listings of fusible alloy compositions and melting points are available.[5] As a point of reference, paper will spontaneously begin to burn when heated to 450°F, just above the melting point of tin.

Table 6.1

Melting Points of Fusible Alloys of Variable Composition: The Melting Points of Several Alloys of Varying Metal Composition That Are Employed in the Heat-Activating Mechanism of an Automatic Fire Sprinkler Head

Bi	Cd	In	Pb	Sn	Melting point/°C	Melting point/°F
45	5	19	23	8	47	117
49		21	18	12	58	136
50	10		27	13	70	158
42	9		38	11	70–88	158–190
52			32	16	95	203
55			45		124	255
58				42	138	280
40				60	138–170	280–338

KEY TERMS

alloy

REFERENCES

1. "Reliable Model ZX-QR-INST Pendent Horizontal Sidewall Quick Response Institutional Sprinkler," The Reliable Automatic Sprinkler Co., Inc., http://www.reliablesprinkler.com/sprinklr/137c.htm
2. "Reliable Model F1 and Model F1FR Intermediate Level Sprinklers," The Reliable Automatic Sprinkler Co., Inc., http://www.reliablesprinkler.com/sprinklr/170b.htm
3. "Sprinkler Systems," on the Fireworld Web Site for Firefighters, http://www.fireworld.net
4. "Fusible Alloys," Clad Metal Industries, http://www.cladmetalindustries.com/Alloys.html
5. "Fusible Alloys with Characteristics," Canfield Technologies, http://www.solders.com/aloyprod.htm

RELATED WEB SITES

Artim, N. (1994, September). An introduction to automatic fire sprinklers, Part I. *WAAC Newsletter,* **16,** no. 3, 20, http://sul-server-2.stanford.edu/waac/wn/wn16/wn16-3/wn16-309.html
Artim, N. (1995, May). An introduction to automatic fire sprinklers, Part II. *WAAC Newsletter,* **17,** no. 2, http://palimpsest.stanford.edu/waac/wn/wn17/wn17-2/wn17-206.html
"Component Descriptions: 1.3 Fire Detection Devices. 1.3.2 Thermal Detectors," The Reliable Automatic Sprinkler Co., Inc., http://www.reliablesprinkler.com/hazards/707-1.htm#Page21
"Composition of Alloys," Oliver Seely, professor of chemistry, California State University, Dominguez Hills, http://155.135.31.26/oliver/chemdata/alloys.htm
"Glass Bulb Sprinkers," Star Sprinkler, Inc., http://www.descartes.net/star/products/sprinklers/glassbulbsprinklers.html

Chapter 7

Connections to the Household

WHAT IS THE DARK SPOT ON THE INSIDE OF A LIGHTBULB WHEN IT BURNS OUT?

> "We are striking it big in the electric light,
> better than my vivid imagination first conceived.
> Where this thing is going to stop Lord only knows."[1]
> —Thomas Edison, October 1879

In his lifetime, Thomas Alva Edison patented 1093 inventions, but the incandescent lightbulb is generally regarded as his most famous invention. On October 21, 1879, 29-year-old Thomas Edison demonstrated the first incandescent lamp in Menlo Park, New Jersey. The bulb burned for 13-1/2 hours. The chemistry of the components of this lighting device are essential to its incredible success.

THE CHEMICAL ESSENCE

The dark spot on the inside of a burned-out lightbulb is tungsten metal that has sublimed (vaporized from the solid) as a consequence of the heating of the tungsten filament to produce white light.

THE CHEMICAL SPECIFICS

Incandescent lamps consist of glass bulbs that enclose an electrically heated filament that emits light. For more than 50 years prior to Thomas Edison's success, scientists had experimented with developing electric lamps. With financiers such as J. P. Morgan and the Vanderbilts, Edison founded the Edison Electric Light Company in 1878. Its prime mission was to generate cheap electric power to provide an illumination source. For the filament of his electric lamp, Edison reportedly[1] experimented on 6000 different types of materials, eventually narrowing his focus on fine platinum wire and a mix of 10% iridium with platinum. Unfortunately, these filament choices were unsuccessful, for the materials would not handle the current without melting. Edison's final (and successful) incandescent lamp utilized carbonized cotton filaments.

Modern incandescent lamps employ tungsten filaments. What properties of a material enhance its use as a filament in incandescent lightbulbs? Materials with a high resistance to the flow of electricity and a high melting temperature are ideal, for the resistance will cause heat to be generated in the material until it glows white. A high melting temperature ensures that the glow is maintained. Because hot metals emit only small amounts of light in the infrared range, a metal filament maximizes the amount of visible light emitted and minimizes infrared radiation that would generate heat.[2]

What physical and chemical properties of tungsten make this metal an ideal choice for the filament material? As a practical consideration, the ductility of tungsten enables the production of filaments that consist of fine wires. Furthermore, because an electric current generates high temperatures, a substance like tungsten with a high melting point permits a longer lifetime of incandescence. In fact, tungsten has the highest melting point of any element − 3683 K or 3410°C or 6170°F.[2] As lightbulbs are often filled with a gas (argon and nitrogen, for example) to carry heat away from the filament, the filament material should be chemically inert with respect to these gases. Tungsten meets this criterion, exhibiting no chemical reactivity with these gases. By keeping the tungsten filament "cooler" with the presence of inert gas, less tungsten is lost by sublimation. Otherwise, the thinning of the filament by the sublimation loss of tungsten would increase the resistance to current flow, thereby increasing the filament temperature. Eventually, the resistance would reach a high enough level that the current required to produce light would essentially vaporize the tungsten metal. As an additional benefit, the increased pressure in the system due to the added gas also aids in reducing the extent of sublimation of the filament.

"Long-life" incandescent lightbulbs use different combinations of gases to fill the bulb in order to lengthen the life of the tungsten filament.

In contrast to normal lamps filled with inert gases, halogen lights are filled with gaseous iodine or bromine to take advantage of the chemical reactivity of tungsten and gaseous halogens. At the high temperatures (>3000°C) near the tungsten filament in an operating halogen lamp[3], tungsten does not combine chemically with these halogens. However, nearer the wall of the bulb where temperatures may be 800 to 1000°C cooler,[3] a gaseous tungsten halide forms. This vapor eventually migrates to the filament and decomposes at high filament temperatures, depositing tungsten metal on the glowing filament and regenerating the halogen gas. This ongoing cycle of chemical reaction and chemical decomposition to maintain the tungsten filament yields a brighter and longer-lasting light.

KEY TERMS

incandescence

REFERENCES

1. "Incandescent Lamp," Edison Internet Museum, http://www.naples.net/~nfn04538/bulb_1.htm
2. "Electromagnetic Radiation: General Considerations: Generation of Electromagnetic Radiation: Continuous Spectra of Electromagnetic Radiation," *Encyclopedia Britannica Online*, http://www.eb.com:180/cgi-bin/g?DocF=macro/5002/7/3.html
3. "*Scientific American*: Working Knowledge: Halogen Lamps," Terry McGowan, July 1996, http://www.sciam.com/0796issue/0796working.html

RELATED WEB SITES

"A Biography of Thomas Alva Edison," The Academy for the Advancement of Science and Technology, http://www.bergen.org/AAST/Projects/Timeline/Housing19/historyl.htm

"Edison Birthplace Museum-Related Resources," http://www.tomedison.org/ref.html

"Incandescent Filament," SCE&G, A Scana Company, http://www.scana.com/sce%26g/business_solutions/lighting/linfila.htm

"Incandescent Lighting," SCE&G, A Scana Company, http://www.scana.com/sce%26g/business_solutions/lighting/linbfin.htm

McAuliffe, K. (1995, December). The undiscovered world of Thomas Edison. *The Atlantic Monthly*, **276,** no. 6, 80–93, http://www.theAtlantic.com/atlantic/issues/95dec/edison/edison.htm

"Thomas A. Edison Papers," Rutgers, The State University of New Jersey, http://edison.
rutgers.edu:80/taep.htm

HOW ARE LIGHTBULBS FROSTED?

*In 1925 Kyozo Fuwa (Toshiba) developed the first frosted glass lightbulb,
generally regarded as one of the five greatest lightbulb inventions.[1] The softer
light created by a frosted lightbulb is accomplished through a chemical
reaction with the inner surface of the bulb.*

THE CHEMICAL ESSENCE

The interior of a "frosted" lightbulb is etched with hydrofluoric acid,
HF. A chemical reaction between the glass and the acid produce a white
substance that coats the bulb's interior surface. Frosted lightbulbs were
introduced in 1925 to provide a diffused light rather than the glaring light
of an unconcealed filament.[1] The decorative design of etched glass is also
accomplished through the corrosive action of hydrofluoric acid. Typically,
glass is coated with layers of paraffin or beeswax through which patterns
are traced with metal needles. Dipping the glass in an aqueous solution of
HF etches the design in the unprotected surface.

THE CHEMICAL SPECIFICS

Hydrofluoric acid reacts with glass via an overall reaction that may be
summarized as follows:

$$SiO_2 \text{ (s)} + 6 \text{ HF (aq)} \rightarrow H_2SiF_6 \text{ (aq)} + 2 \text{ H}_2O \text{ (l)}$$

(Recall that, because of the strong $H-F$ bond, hydrofluoric acid is a weak
acid with a small acid dissociation constant K_a of 6.8×10^{-4}. In contrast,
the other binary acids of the halogen family—HCl, HBr, and HI—are
strong acids that completely dissociate in water.) The fluorosilicic acid
produced, H_2SiF_6, is a water-soluble substance with a structure as in Figure 1.

As we know, glass is defined as an extended three-dimensional network
of atoms that forms a solid lacking the orderly arrangement or long-range
periodicity of a crystalline material. Despite the amorphous nature of glass,
a definite chemical composition may be ascribed to the material. Most
commercially important glasses are silica glasses with SiO_4 tetrahedra as

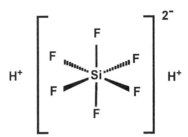

Figure 1 The chemical structure of fluorosilicic acid, produced by the action of hydrofluoric acid (HF) on glass.

the network building blocks. The network-forming Si^{4+} silicon ions are bonded to four oxygen atoms. These oxygen atoms fall into two classifications, depending on the oxide content of the glass. A bridging oxygen atom connects two SiO_4 tetrahedra, whereas a nonbridging oxygen atom is linked with a "network-modifying" cation such as a sodium ion. Silica glass (also known as vitreous silica) contains 100% bridging oxygen atoms. A variety of network-modifying cations are present in soda-lime silicate glass (with $O-Na$, $O-Ca$, $O-Al$, and $O-Mg$ linkages), sodium borosilicate glass (with $O-Na$, $O-Al$, and $O-B$ linkages), and aluminosilicate glass (with linkages of oxygen to Na, Ca, Al, Mg, and B possible). Many properties of the glass—density, viscosity, chemical durability, electrical resitivity— are related to the "connectivity" of the structure, (i.e., the concentration of nonbridging oxygens), and the nature of the network-modifying ions.[2] In particular, the chemical resistance of glass is enhanced by increasing the silica content at the surface. Silica glass offers the greatest resistance. Normal weathering of glass occurs via an ion-exchange process whereby alkali ions in the glass are exchanged with hydronium ions present in water or atmospheric humidity. Further depositing of white alkali carbonates and bicarbonates on the glass surface can occur as the exchanged alkali ions react with atmospheric carbon dioxide and water. Dissolution of the entire silica network is possible when hydrofluoric acid (as well as perchloric and phosphoric acids and caustic alkalis) attack the surface of silicate glasses. The chemical durability of glass can be modified by fire-polishing (which removes alkali ions by volatilization) or by treating the glass surface with a polymeric coating.

KEY TERMS

weak acid

REFERENCES

1. "Thomas Edison 1997 Main Exhibition: Filament & Incandescent Electric Light," Tepia, http://www.tepia.or.jp/main/filainfe.htm
2. "Industrial Glass: Properties of glass," *Encyclopedia Britannica Online,* http://www.eb.com:180/cgi-bin/g?DocF=micro/5009/16/1.html

RELATED WEB SITES

"The Corning Glass Museum: A Resource for Glass: Educational Resources," http://www.pennynet.org/glmuseum/
"Lighting Systems: Bulb Finish," SCE&G, A Scana Company, http://www.scana.com/sce%26g/business_solutions/lighting/linbfin.htm

WHAT MAKES A LAVA LAMP WORK?

Many fundamental chemical principles underlie the operation of this intriguing novelty device.

THE CHEMICAL ESSENCE

The lava lamp has captivated viewers for more than 50 years. Shortly after discovering the lava lamp's prototype—a "contraption made out of a cocktail shaker, old tins, and things"—in a Hampshire, England, pub after World War II, Edward Craven Walker founded the Crestworth Company in Dorset, England, to develop the device.[1] Over a 15-year period he perfected the lava lamp and mass marketed the light source. Although the demand for lava lamps reached fad proportions in the 1960s, a resurgence of the lava lamp craze occurred in the 1990s. Walker's company, sold in 1990, still produces lava lamps, and Haggerty Enterprises of Chicago produces Lava Lites in the United States. A vast number of color combinations exist.

Two liquids with vastly different properties are the essential ingredients of a lava lamp. The two liquids must differ significantly in chemical composition to prevent mixing. The liquid that constitutes the "lava" is generally a substance that does not dissolve readily in water (i.e., a nonwater soluble substance). The second liquid is a water-soluble material or a mixture of water and a liquid that dissolves readily in water. The inability of the "lava" and the water-based solution to mix encourages the lava to coalesce to form the familiar "blobs" that migrate through the water-based liquid. The lava material must be slightly more dense than water at room temperature, positioning the lava at the bottom of the container before the lamp is switched on. An illuminated lamp, usually a 40-watt bulb, serves as a heat source.[2] A small coil of wire at the base of the lamp also serves as a heating element within the liquid compartment. Both liquids decrease in density as the temperature of the system increases, but the liquids are selected so that the lava liquid changes density more dramatically than the water solution as the temperature rises. The now less dense ("lighter") lava rises in the container and also expands in volume due to higher temperatures. A liquid with a moderate viscosity contributes to the motion and circulation of the lava. As the lava moves away from the heat source, it begins to cool, decreasing its volume (contracting the "blobs") and increasing its density (sinking to the bottom of the lamp). The cycle repeats to generate the mesmerizing and ever-changing pattern formation and motion of the lava. Different colored dyes—one that dissolves readily in a water-based mixture and a second that disperses only in the lava liquid—are added to produce the various color combinations.

THE CHEMICAL SPECIFICS

Liquids with disparate intermolecular forces are critical for the operation of a lava lamp. The patent issued in 1971[1] specifies mineral oil (a hydrocarbon-based liquid) and a 70/30% by volume water/propylene glycol mix as the insoluble liquids. Other water-insoluble substances with specific gravities greater than that of water (i.e., densities greater than water) and with larger thermal coefficients of expansion (i.e., greater temperature-dependent densities to increase liquid volume upon heating) include benzyl alcohol, cinnamyl alcohol, diethyl phthalate, and ethyl salicylate. Isopropyl alcohol, ethylene glycol, and glycerol are water-soluble substitutes for propylene glycol. The hydrophobic forces that dominate in the lava liquids are incompatible with the hydrogen bonds and dipole interactions that exist in the water-based mixtures. Because the forces of attraction to produce a single liquid phase are not present, two distinct phases emerge.

Six patents that reference Craven's original patent have been issued. One of the earlier patents[3] details a three-phase liquid system to create a "device for producing aesthetic effects." The patent specifications involve at least two liquids of different density that are not completely or permanently miscible with each other. In particular the three liquids *A, B, C,* denoted in order of increasing density, are described as follows: "*A* liquid paraffin and/or silicone oil and/or naphthene and/or hexachlorobutadiene; *B* water or an ether, more particularly propanetrioxyethylether or a polyether; *C* esters with chemically bound phosphorus and/or chemically bound halogen, preferably chlorine, an ester of phthalic acid, and more particularly dibutoxybutyl phthalate, a carbonic acid ester, more particularly propanediol carbonate, or ethanediolmonophenylether or tetrahydrothiophene-1,1-dioxide with the provision that the selected liquids are not completely and not permanently miscible with each other".[4] In a later patent (US#4419283: 12/06/1983, "Liquid compositions for display devices"),[5] an invention is described consisting of systems of three, four, and five mutually immiscible liquid phases that are suitable for use in display devices and novelty toys. The preferred four-phase systems are comprised of one highly hydrophobic organic phase, one organic phase containing compounds that are moderately polar, one phase containing hydrogen-bonding organic compounds, and one aqueous phase. Numerous examples of specific liquids that fulfill these conditions are delineated. As long as there is interest in such novelty devices as lava lamps, the search continues for room temperature, mutually immiscible liquid systems comprised of water and organic liquids that are inexpensive, nontoxic, noncombustible, and readily dyeable!

KEY TERMS

miscibility; solubility; density

REFERENCES

1. "Lava Lite Lamp," [US Patent 3,570,156 March 16, 1971], *Your Mining Co. Guide to Inventors;* http://inventors.miningco.com/library/weekly/aa092297.htm
2. Hubscher, R. (1991, March). How to make a lava lamp. *Popular Electronics,* 31 (4); also in Popular Electronics' *1992 Electronics Hobbyists Handbook*
3. These patents are U.S. Patent #5778576, issued on 07/14/1998, "Novelty lamp," #5709454: 01/20/1998, "Vehicle visual display devices"; #D339295: 09/14/1993, "Combined bottle and cap"; #4419283: 12/06/1983, "Liquid compositions for display devices"; #4085533:04/25/1978, "Device for producing aesthetic effects"; #4034493: 07/12/1977, "Fluid novelty device."
4. "4085533: Device for producing aesthetic effects," http://www.patents.ibm.com/details?patent_number=4085533
5. "4419283: Liquid compositions for display devices," IBM US Patent Database, http://www.patents.ibm.com/details?patent_number=4419283

RELATED WEB SITES

"The Perpetual Motion of the Lava Lamp: A PY105 Physics Fair Project—by Andrew Bazylevsky," http://acs6.bu.edu:8001/~uke/lava.html
"Self-Made Lavalamps," Http://www.forwiss.tu-muenchen.de/~fenk/public/lavalamps/Lavalamps.html

WHY SHOULDN'T AMMONIA CLEANSERS BE MIXED WITH BLEACH?

Both ammonia and bleach are useful household products for cleaning stains, sanitizing surfaces, disinfecting, and deodorizing. But the chemistry of these individual products dictates that you always heed the warning: **Caution: Never mix bleach and ammonia cleansers.**

THE CHEMICAL ESSENCE

Liquid household bleach is generally a 5% solution of sodium hypochlorite ($NaOCl$). Ammonia cleansers—including general household cleansers, wax removers, glass and window cleaners, and oven cleaners—are aqueous solutions of 5 to 10% ammonia, NH_3. Mixing bleach with cleansers containing ammonia leads to the formation of a family of potentially toxic compounds known as *chloramines*. These toxic gases have acrid fumes that can burn mucous membranes. Scented bleaches can mask one's natural ability to detect these harmful fumes.

THE CHEMICAL SPECIFICS

Ammonia (NH_3) and hypochlorite ion (OCl^-) combine to produce three different chloramine species—that is, compounds that are derivatives of ammonia in which one or more of the hydrogen atoms has been replaced by a chlorine atom. In order of increasing degree of chlorine substitution, these chloramines are named monochloramine (NH_2Cl), dichloramine ($NHCl_2$), and nitrogen trichloride (NCl_3).

$$OCl^- + NH_3 \rightarrow OH^- + NH_2Cl$$

$$OCl^- + NH_2Cl \rightarrow OH^- + NHCl_2$$

$$OCl^- + NHCl_2 \rightarrow OH^- + NCl_3$$

The pungent and irritating odor of chloramines is often mistaken for the chlorine odor of swimming pools. Chloramines form from the combination of sodium hypochlorite (added to sterilize the water) and nitrogen-containing compounds that are human waste by-products.

In the presence of excess ammonia, hypochlorite ion and ammonia can combine to form hydrazine (N_2H_4), another toxic and potentially explosive substance:

$$OCl^- + NH_3 \rightarrow NH_2Cl + OH^-$$

$$NH_2Cl + NH_3 + OH^- \rightarrow N_2H_4 + Cl^- + H_2O$$

Industrial preparation of hydrazine is based on this reaction of ammonia with an alkaline solution of sodium hypochlorite, known as the Raschig process introduced in 1907.

KEY TERMS

chloramines; hydrazine

RELATED WEB SITES

"Pool Water Chemistry: Technical Details," Virtual Pool & Spa Store, Long Island Hot Tubs & Paramount Pools, http://paramountpools.com/page371.htm

WHY CAN FLOOR WAXES BE REMOVED EASILY WITH AMMONIA CLEANSERS?

"Ultra Heavy Duty, High Grade floor finish based on the advanced technology of interlocking metal polymers."[1] "Superior innovation, incorporating advanced polymer technology, provides you the highest quality floor finishes."[2] How has chemistry improved the modern floor finish?

THE CHEMICAL ESSENCE

Modern floor finishes contain mixtures of ingredients that when applied as a liquid dry to give a clear, durable film coating. Most floor finishes are combinations of polymers blended to provide unique and desirable characteristics such as durability, gloss retention, resistance to scuffing, fast drying, and ease of removal. Polymers made up of more than one monomer or repeating unit are called "copolymers."

Polymeric surface coatings often gain their strength and durability from a three-dimensional network structure created by a cross-linking of the polymer chains. It is this network structure that can give floor waxes their unique properties. New formulations use metal ions to create the cross-linking structure. These metal ions are often soluble in ammonia cleansers, destroying the polymeric structure and allowing the floor wax to be stripped.

THE CHEMICAL SPECIFICS

The most common polymer groupings in floor waxes are acrylics, acrylic copolymers, and urethanes. An acrylic polymer is a generic term denoting derivatives of acrylic acid ($CH_2=CHCOOH$) and methacrylic acid ($CH_2=C(CH_3)COOH$), including acrylic esters ($CH_2=CHCO_2R$), and compounds containing nitrile ($-CN$) and amide ($-CONH_2$) groups such as acrylonitrile ($CH2=CHCN$) and acrylamide ($CH_2=CHCONH_2$). A polyurethane is a polymer formed from linear repetitions of the monomer urethane ($CH_3CH_3OCONH_2$).

One recent development is the "metal interlock floor finish," a fast-drying, durable, yet easily removed finish.[3,4] The floor wax formulations contain zinc[5], often in the form of the transition metal complex ion $Zn(NH_3)_4{}^{2+}$. As the liquid dries, the zinc ion is released upon evaporation of ammonia:

$$Zn(NH_3)_4{}^{2+} \text{ (aq)} \rightarrow Zn^{2+} \text{ (aq)} + 4\ NH_3 \text{ (g)}$$

The zinc ion cross-links with the polymer to create sufficient strength and cross-linking density for durability and resistance to abrasion and detergents. The floor finish is easily removed with ammonia cleansers to reform the stable complex with zinc.[6] The action of pulling the zinc out of the polymer allows the polymer to dissolve in the stripping solution.

KEY TERMS

transition metal complex ion; copolymers; cross-linked polymers

REFERENCES

1. "Phoenix—30% Premium Sealer & Finish," Space Chemical, Inc., http://www.microserve. net/~spaceinc/floor.html
2. "Resilient Floor Care," Betco Products, http://www.betco.com/allprod/navprod.htm
3. Spartan Chemical Floor Finishes, http://www.spartanchemical.com
4. "Metal Lock—Technical Data Sheet," Ameqex, Inc., http://www.iswt.com/ameqex/ liquids/r860.htm
5. "So what's the real problem with Zinc?" Tom Wright, Technical Director of Betco Corporation, 1991, http://www.michco.com/HelpStuff/Zinc_Problem_Help.html
6. "Floor Care Products—High pH Cleaners," Ameqex, Inc., http://www.iswt.com/ameqex/ liquids/floorc.htm

RELATED WEB SITES

"Update: Floor Care Chemicals," Glen Franklin, Maintenance Solutions, May 1997, http://www.cleanlink.com/NR/NR2m7ea.html

WHY DO MANUFACTURERS RECOMMEND THAT CONSUMERS CLEAN THEIR AUTOMATIC COFFEE MAKERS WITH VINEGAR OR RUN VINEGAR THROUGH THEIR STEAM IRONS?

Vinegar has been called Mother Nature's Liquid Gold. At the very least, vinegar is recognized as a "natural" and effective household cleaner. Why is vinegar recommended for cleaning automatic coffee makers and steam irons?

THE CHEMICAL ESSENCE

Vinegar is recommended for cleaning a variety of appliances and other items that may be stained by hard-water deposits. Automatic coffee makers, steam irons, dishwashers, teapots, faucet heads, and shower heads all, over time, accumulate calcium deposits from hard water. Groundwater—that is, water that travels through soil and rocks—accumulates dissolved calcium ions as a consequence of the natural weathering of minerals that contain calcium, such as limestone, calcite, shells, and coral. At the same time, carbon dioxide in the air dissolves in water to form carbonate ions that combine with calcium ions to form a white solid, calcium carbonate. Vinegar removes the white deposits of calcium carbonate by dissolving the solid.

THE CHEMICAL SPECIFICS

Vinegar or acetic acid combines with calcium carbonate to dissolve the precipitate, form free calcium ions and water, and liberate carbon dioxide gas:

$$2 \ CH_3COOH(aq) + CaCO_3(s) \rightarrow Ca^{2+}(aq) + 2 \ CH_3COO^-(aq) + CO_2(g) + H_2O(l)$$

The dissolution of calcium carbonate by vinegar can also be observed when a hard-boiled egg is placed in a vinegar solution. Overnight the egg shell, composed of calcium carbonate, will dissolve leaving only an outer membrane. Chalk, also composed of calcium carbonate, will also dissolve in a vinegar (or acid) solution. In preparing hard-boiled eggs to be treated with colored dyes, as in dyeing Easter eggs, consumers are often recommended to add a small amount of vinegar to the aqueous dye solution. This practice helps the dye adhere to the egg shell by creating a freshsurface on the egg exterior through partial dissolution of the old surface.

KEY TERMS

dissolution; precipitate

WHY DO HOMEMADE RECIPES FOR COPPER CLEANER CALL FOR VINEGAR?

The blue-green coating or "patina" on many bronze statues and copper artifacts often adds to the authenticity of the art. Less desirable are the dark green and black coatings that develop on cooking utensils or jewelry items. How does understanding the chemistry of the copper corrosion layer enable you to devise a homemade recipe for cleaning?

THE CHEMICAL ESSENCE

One simple yet successful technique for cleaning green, faded copper utensils, brass items, and chrome surfaces recommends the use of vinegar: "Pour vinegar and salt over copper surface and rub." Even black- or green-coated specimens of native copper can be cleaned with vinegar. Alternative recipes suggest lemon juice, ketchup, or even water in which onions have been boiled. Why do these remedies work?

Copper is a metallic element; brass is an alloy or mixture of the metallic elements copper and zinc. The surfaces of copper and brass items tarnish with prolonged exposure to air, particularly in moist environments with high carbon dioxide (CO_2) or sulfur dioxide (SO_2) concentrations. The compounds that form on the surface, ranging in color from black to blue to dark green, dissolve readily in acidic solutions. Vinegar contains acetic acid, ketchup contains tomatoes rich in ascorbic acid (vitamin C), and onions contain malic acid and citric acid. All of these foods provide variable amounts of acid to dissolve the tarnish on copper surfaces.

THE CHEMICAL SPECIFICS

What is the identity of the corrosion layer (called *patina*) on copper metal surfaces? The oxidation of copper metal can be induced by a number of substances present in our atmosphere, including oxygen, carbon dioxide, sulfur dioxide, and hydrogen sulfide. Copper oxides, sulfate and carbonate salts of copper, and copper sulfides result. In seacoast locations, chloride salts may form an essential part of the patina film. Some specific copper oxidation products include green malachite $CuCO_3 \cdot Cu(OH)_2$, blue azurite $2CuCO_3 \cdot Cu(OH)_2$, chalcocite Cu_2S, covellite CuS, chalcopyrite $CuS \cdot FeS$, bornite $FeS \cdot 2Cu_2S \cdot CuS$, tetrahedrite $4Cu_2S \cdot Sb_2S_3$, brochantite $(CuSO_4 \cdot 3Cu(OH)_2$, and atacamite $(CuCl_2 \cdot 3Cu(OH)_2$.[1] The green and blue colored patinas are generally composed of cupric carbonates, including green malachite and blue azurite. Other green coatings are generally combinations of copper sulfate (brochantite) and copper chloride (atacamite). Black coatings on copper are cupric oxide (copper (II) oxide or the mineral known as tenorite, CuO), whereas red corrosion layers arise from cuprous oxide (copper (I) oxide or cuprite, Cu_2O).

Recently the patina formed in the atmosphere on the roof of the Stockholm City Hall was analyzed.[2] Several components of the patina were identified, including brochantite $(CuSO_4 \cdot 3Cu(OH)_2)$, antlerite $(Cu_3(OH)_4 SO_4)$, and basic cupric carbonate $(Cu_2CO_3(OH)_6H_2O)$. At the Rodin Museum in Philadelphia, active corrosion of Rodin's *The Thinker* was ascribed in 1992 to primarily brochantite $(CuSO_4 \cdot 3Cu(OH)_2)$, the basic copper sulfate.[3] This corrosion was attributed to an "acid rain" atmosphere as a consequence of the display of the sculpture in an outdoor urban-industrial environment for more than 60 years. During the major restoration of the Statue of Liberty in 1986, the variable blue discoloration on the copper skin of the surface was attributed to a number of minerals. A phase diagram was constructed to assign the colors on the statue's surface to particular minerals that had formed according to the varying exposure to sea spray, rainwater, and sulfur dioxide air pollution.[4] The various blue shades were assigned to brochantite $(CuSO_4 \cdot 3Cu(OH)_2$, antlerite $(Cu_3(OH)_4SO_4)$, and chalcanthite or hydrated copper sulfate $(CuSO_4 \cdot 5H_2O)$ and reddish coloring was ascribed to cuprous oxide (copper (I) oxide or the mineral known as tenorite). These analyses clearly demonstrate that the exposure of copper metal to the atmosphere results in complex layers of minerals containing oxides, hydroxides, sulfates, carbonates, chlorides, and sulfides.

KEY TERMS

oxidation; alloy

REFERENCES

1. "Copper Corrosion Study," F. Romero, California State University—Dominguez Hills, http://www.csudh.edu/math/fromero/chemistry/copper/copper.htm
2. Alex O. Salnick and Werner Faubel,"Photoacoustic FT-IR Spectroscopy of Natural Copper Patina," Photothermal and Optoelectronic Diagnostics Laboratory, University of Toronto, Department of Mechnical Engineering, Toronto, Ontario M5S 1A4, Canada, *Appl. Spec.* v49 (10) (1998)
3. "The Conservation of Rodin's *The Thinker,*" Philadelphia Museum of Art, http://www.philamuseum.org/resources/conservation/projects/rodin/3.shtml
4. "Transferring Technology from Conservation Science to Infrastructure Renewal," Richard A. Livingston, http://www.tfhrc.gov/pubrds/summer94/p94su18.htm

RELATED WEB SITES

"Cleaning Native Copper," Herb Sulsky, *Lithosphere,* May 1993, Fallbrook Gem and Mineral Society, http://www.inetworld.net/rbusch/fgms/copper.htm

"Conservation of Cupreous Metals (Copper, Bronze, Brass)," http://nautarch.tamu.edu/class/anth605/File12.htm

"Conservation of Nonferrous Metals," http://128.174.5.51/denix/Public/ES-Programs/Conservation/Underwater/5-CU-AG.html

"Copper: The Red Metal," The Geology Project Homepage, Tom Lugaski, University of Nevada, Reno, http://www.unr.edu/sb204/geology/coptext.html

"Formula for Copper & Brass Cleaner," http://www.makestuff.com/brass_copper.html

"The Mineral Antlerite," The Mineral and Gemstone Kingdom, http://mineral.galleries.com/minerals/sulfates/antlerit/antlerit.htm

"The Mineral Brochantite," The Mineral and Gemstone Kingdom, http://www.minerals.net/mineral/sulfates/brochant/brochant.htm

"The Restoration and Conservation of Ancient Copper Coins," Doyle W. Lynch, Baylor University, http://www2.dcci.com/dlynch/digbible/restoration.shtml

WHY DOES BAKING SODA EXTINGUISH A FIRE?

A box of baking soda is often recommended as a handy fire extinguisher in the kitchen. This household hint takes advantage of a chemical reaction involving baking soda as a reactant to douse the fire.

THE CHEMICAL ESSENCE

Three key components are required to sustain a fire: a fuel (e.g., a hydrocarbon compound such as methane or octane or a piece of wood containing cellulose), an oxidizing agent (generally supplied by oxygen in air), and heat. Combustion of the fuel is terminated when the supply of either the fuel, the oxidizing agent, or the heat is eliminated. Many fire-fighting techniques focus on removing the last two ingredients. For example, fires are often smothered by restricting the flow of oxygen to the fuel, as in the action of extinguishing a flame by throwing a fire blanket over the burning object. Similarly, the oxygen atmosphere can be replaced by a gas that will not support combustion, such as carbon dioxide (CO_2). The action of CO_2 fire extinguishers relies on the fact that CO_2 is heavier than air (44.0 g mol^{-1} versus 28.9 g mol^{-1})[1], enabling a CO_2-enriched environment to settle over and envelop the burning fuel. The displacement of air by the denser CO_2 gas terminates the combustion process. Finally, water battles fires by cooling the system and lowering the temperature to a point where the reaction known as burning ("combustion") can no longer be sustained.

Solid sodium bicarbonate, $NaHCO_3$, otherwise known as baking soda, is also an excellent fire extinguisher because the chemistry of this salt at high temperatures diminishes the oxygen environment. How does this occur? Although $NaHCO_3$ is a stable ionic solid at room temperature, at the high temperatures typical of fires $NaHCO_3$ (s) undergoes *thermal decomposition* (i.e., a heat-activated reaction) to produce carbon dioxide gas as one of the by-products. Just as in the case of CO_2 fire extinguishers, the CO_2 gas produced by the decomposition of baking soda creates an atmosphere of reduced oxygen content that aids in extinguishing a fire.

The Chemical Specifics

A *thermal decomposition reaction* is a reaction that is activated by heat or high temperatures and that generates simpler (i.e., containing fewer atoms and thus characterized by lower molecular weights) substances from a single complex substance. The overall balanced equation for the thermal decomposition of sodium bicarbonate reveals the simpler substances produced:

$$2 \; NaHCO_3(s) \rightarrow Na_2CO_3(s) + H_2O(g) + CO_2(g)$$

In addition to carbon dioxide gas and water vapor, a white crystalline solid, sodium carbonate, is formed. How does heat induce this reaction? The requirement of an input of heat to decompose the ionic lattice of sodium bicarbonate is supported by a calculation[2] of the standard enthalpy change for this reaction at 298 K, an *endothermic* quantity of $+135.6$ kJ mol^{-1}, meaning heat is required for the reaction to occur. The overall balanced equation also clearly depicts the heterogeneous nature of the decomposition process. A *heterogeneous reaction* is characterized by the presence of multiple phases in the reaction mixture; here both solid and gas phases exist at the high temperatures at which this reaction occurs. One of the driving forces for this reaction to go to completion is the loss of the gas products to the atmosphere. Thus, the chemistry of sodium bicarbonate at elevated temperatures is a convenient source of carbon dioxide to eliminate the flow of oxygen to a burning fuel.

KEY TERMS

combustion; endothermic reaction; heterogeneous reaction; thermal decomposition reaction; fuel; oxidizing agent

REFERENCES

1. Using a composition for dry air of approximately 78.1% $N_2(g)$, 21.0% $O_2(g)$, and 0.9% $Ar(g)$, an average molecular weight for air is estimated at 28.9 g mol^{-1}.
2. Using standard molar enthalpies of formation at 298 K for $NaHCO_3(s)$, $Na_2CO_3(s)$, $H_2O(g)$, and $CO_2(g)$ of -950.8 kJ mol^{-1}, -1130.7 kJ mol^{-1}, -241.8 kJ mol^{-1}, and -393.5 kJ mol^{-1}, respectively. [Lide, D. R. (Ed.) (1993). Standard Thermodynamic Properties of Chemical Substances. *In: Handbook of Chemistry and Physics*, 74th ed., pp. 5-4–5-47. CRC Press, Boca Raton, FL.]

RELATED WEB SITES

"Arm & Hammer Baking Soda: The Everyday Miracle & Trade," http://www.armhammer. com/whatis.htm

WHY SHOULD ONE NEVER ATTEMPT TO EXTINGUISH A MAGNESIUM FIRE WITH EITHER WATER OR A CO₂ FIRE EXTINGUISHER?

Magnesium blocks with a flint rod molded into the block are sold as emergency fire starters. When the magnesium is shaved into very small pieces, the shavings readily ignite with a flint spark to generate an intense (5400°F) fire. The temperatures are sufficient to ignite even damp wood. Although the shavings burn rapidly and thereby disappear, what chemical principles should you understand to douse a fire involving larger amounts of magnesium?

THE CHEMICAL ESSENCE

Magnesium metal burns vigorously in oxygen to produce a brilliant white flame. One common example of the combustion of magnesium and oxygen is a flashbulb—the intense flash is produced by the reaction of magnesium ribbon with oxygen. But other atmospheres support the burning of magnesium, including nitrogen, carbon dioxide, and water. Some of these reactions are even more vigorous than the reaction of magnesium and oxygen, generating heat and further exacerbating the situation.

Combustible metals, such as magnesium, titanium, zirconium, potassium, and sodium, generate what are classified as Class D fires. These materials burn at high temperatures and react violently with water or other chemicals. Such materials should be handled with care. To extinguish fires involving these flammable metals, metal/sand extinguishers should be used. These extinguishers work by simply smothering the fire.

THE CHEMICAL SPECIFICS

The overall balanced chemical equations for the reaction of solid magnesium with oxygen gas, nitrogen gas, liquid water, and carbon dioxide gas follow. In each reaction, magnesium undergoes an oxidation process (i.e., an increase in oxidation number). In addition, each of these reactions is an *exothermic* process, releasing sizable amounts of heat at constant pressure. The heat generated by these reactions continues to fuel the combustion of magnesium, intensifying the fire.

$$2 \text{ Mg (s)} + O_2 \text{ (g)} \rightarrow 2 \text{ MgO (s)} \ \Delta H_{rxtn}° = -1203 \text{ kJ}$$

$$3 \text{ Mg (s)} + N_2 \text{ (g)} \rightarrow Mg_3N_2 \text{ (s)} \ \Delta H_{rxtn}° = -461 \text{ kJ}$$

$$\text{Mg (s)} + 2 \text{ H}_2\text{O (l)} \rightarrow \text{Mg(OH)}_2 \text{ (aq)} + H_2 \text{ (g)} \ \Delta H_{rxtn}° = -355 \text{ kJ}$$

$$2 \text{ Mg (s)} + CO_2 \text{ (g)} \rightarrow 2 \text{ MgO (s)} + C \text{ (s)} \ \Delta H_{rxtn}° = -810. \text{ kJ}$$

KEY TERMS

combustion; oxidation; exothermic reaction

WHY IS HYDROGEN PEROXIDE STORED IN DARK-COLORED PLASTIC BOTTLES?

In addition to protection from breakage, the brown plastic bottles in which hydrogen peroxide is commonly sold enhance the shelf-life of this product.

Packaging plays an important role in limiting the light-induced chemical reactions of hydrogen peroxide.

THE CHEMICAL ESSENCE

Hydrogen peroxide is a major industrial chemical essential to the organic chemicals industry and also useful in environmental treatment of polluted waters. Dilute aqueous solutions of hydrogen peroxide, H_2O_2, are commercially available for uses as a mild antiseptic (a 3% solution, i.e., 3 g of hydrogen peroxide per 100 g of solution) or bleach (a 6% solution). Hydrogen peroxide is the simplest member of the *peroxide* class of compounds, a class of chemical compounds in which two oxygen atoms are linked by a single covalent bond. Hydrogen peroxide decomposes into water and oxygen upon heating or in the presence of trace amounts of metal ions or metal. Even traces of alkali metal ions dissolved from glass can cause this decomposition, and, for this reason, solutions of H_2O_2 are normally stored in wax-coated or plastic bottles.

THE CHEMICAL SPECIFICS

The chemical structure of hydrogen peroxide in Figure 2 depicts the oxygen—oxygen linkage. A skew configuration exists (Fig. 3) with a dihedral angle sensitive to the extent of hydrogen bonding in solution. In the gas phase a dihedral angle of 111.5° exists, whereas a value of 90.2° is measured for crystalline H_2O_2.[1]

The peroxide family of compounds is not only characterized by the oxygen—oxygen single bond but also by an oxidation state for oxygen that is intermediate between the values observed for oxygen in oxygen gas (O_2) and water (H_2O). The unusual −1 oxidation state of oxygen in hydrogen peroxide ensures that H_2O_2 can act as both a reducing agent and an oxidizing

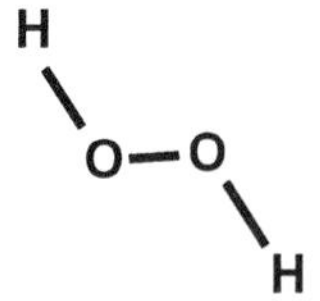

Figure 2 The hydrogen-oxygen-oxygen-hydrogen linkage in hydrogen peroxide.

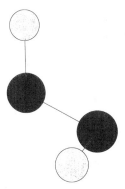

Figure 3   Three-dimensional representation of hydrogen peroxide showing the dihedral angle between the H−O−O and O−O−H planes.

agent. When H_2O_2 acts as a reducing agent, the oxygen atoms are oxidixed from the −1 oxidation state to the 0 oxidation state in O_2:

$$H_2O_2 \text{ (l)} \rightarrow O_2 \text{ (g)} + 2 \text{ H}^+ \text{ (aq)} + 2 \text{ e}^-$$

Alternatively, as an oxidizing agent, H_2O_2 yields H_2O (or OH^- in basic solution) as the oxygen atoms undergo reduction to the −2 oxidation state:

$$H_2O_2 \text{ (l)} + 2 \text{ H}^+ \text{ (aq)} + 2 \text{ e}^- \rightarrow 2 \text{ H}_2O \text{ (l)}$$

Under certain conditions H_2O_2 can simultaneously undergo both oxidation and reduction—that is, *disproportionation*— to yield water and oxygen:

$$2 \text{ H}_2O_2 \text{ (l)} \rightarrow 2 \text{ H}_2O \text{ (l)} + O_2 \text{ (g)}$$

Although this decomposition reaction is slow at room temperature, the disproportionation process can be accelerated by heat, certain catalysts (e.g., metal ions), and light. Traces of alkali metal ions are normally present in aqueous solutions stored in glass bottles, so plastic or wax-coated glass containers reduce the concentration of metal ion catalysts. Brown-color containers limit the wavelengths of light that can be absorbed the solution, restricting the initiation of the disproportionation reaction.

KEY TERMS

disproportionation; peroxide; oxidation; reduction; oxidizing agent; reducing agent; oxidation state

REFERENCES

1. Greenwood, N. N., and Earnshaw, A. (1984). Hydrogen peroxide. *In: Chemistry of the elements,* pp. 744–745. Pergamon Press, Oxford.

RELATED WEB SITES

"Reactive Oxygen Species in Living Systems," Barry Halliwell, http://www.livelinks.com/sumeria/oxy/reactive.html
"Introduction to Hydrogen Peroxide," H_2O_2.com, Southland Environmental and FMC Corporation, http://www.h2o2.com/intro/

HOW DO FURNITURE POLISHES REPEL DUST?

"Dustblock formula actually repels dust." "Cleaning, dusting, and polishing are all completed in seconds." You've heard the claims of various furniture polishes. Is there actually a chemical basis to dusting?

THE CHEMICAL ESSENCE

Furniture polishes claim to dust, clean, shine, and protect wood surfaces. The ingredients within the polish formulation confer these favorable attributes. In particular, to act as a dust repellent, the polish contains antistatic ingredients, that is, hydrocarbon substances that are not prone to static electricity and therefore do not attract dust. Silicone oil[1,2] and lemon oil are typical antistatic ingredients.

Dust particles carry an electrical charge and are attracted to any surface that develops a charge. There are two key ways for a surface to acquire a charge. Friction causes the accumulation of static electricity on surfaces, and thus a surface subjected to frictional forces is susceptible to dust buildup. Additionally, the direct carrying of electrical current by electronic equipment causes such devices to become charged and therefore likely to gather dust. Furniture items attract dust from the charge acquired by frictional forces, and thus the application of an antistatic polish to the furniture surface reduces the likelihood of static electricity. Have you ever wondered why furniture makers discourage "dry dusting," such as dusting without polish? Such action, though perhaps temporarily removing a layer of dust,

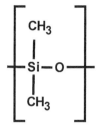

Figure 4   The repeating unit in the most common siloxane polymer, polydimethylsiloxane.

can cause microscopic scratches and even enhance the level of static electricity on the surface.

THE CHEMICAL SPECIFICS

What happens when static charge develops as a consequence of frictional forces between two surfaces? A charge separation occurs, with electrons migrating from one surface to the other. The electron-deficient surface becomes positively charged, while the surface gaining electrons acquires an overall negative charge. Hydrocarbons exhibit weak intermolecular forces—dispersion forces—because of their nonpolar nature and their low *polarizability* (i.e., ease with which an electric field can deform the electronic charge distribution). Antistatic ingredients such as the hydrocarbon-based silicone oil and lemon oil thus deter the creation of friction-induced electrical charges.

Other synonyms for silicone oil, a mixture of many ingredients, include silicone, siloxane, and polydimethylsiloxane. In actuality, siloxanes are a

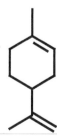

Figure 5   Limonene or methyl-4-isopropenyl-1-cyclohexene, a common ingredient of lemon oil.

diverse class of industrial polymers with a backbone of alternating silicon and oxygen atoms. Organic substituents, commonly methyl ($-CH_3$) groups and phenyl ($-C_6H_5$) rings, are attached to the tetravalent silicon atoms. The repeating unit in the most common siloxane polymer, polydimethyl-siloxane, is given in Figure 4. Lemon oil contains limonene or methyl-4-isopropenyl-1-cyclohexene, as shown in Figure 5.

KEY TERMS

dispersion forces; nonpolar; polarizability; static electricity

REFERENCES

1. "S C Johnson Wax—Pledge—Wax, Spray, Refinish, Aerosol, Material Safety Data Sheet," http://www.biosci.ohio-state.edu/~jsmith/MSDS/PLEDGE.htm
2. "Boyle-Midway—Old English Lemon Cream Furniture Polish, 1865, Material Safety Data Sheet," http://www.biosci.ohio-state.edu/;en;~jsmith/MSDS/OLD%20ENGLISH%20LEMON%20CREAM%20FURNITURE%20POLISH.htm

RELATED WEB SITES

"Introduction to Static Electricity," http://www.simco.nl/static0.htm
"Static Electricity," Electrostatics, Incorporated, http://www.electrostatics.com/page2.html

WHY DOES SOAP SCUM FORM? (OR WHY ARE PHOSPHATES USED IN DETERGENTS?)

Water softener manufacturers remind us of the many undesirable effects produced by hard water. Besides enhancing the likelihood of harmful scale deposits in plumbing, water heaters, and dishwashers, hard water also has been associated with bathtub scum, deposits on laundry, scale on glasses and dishes, scratchy skin, and unmanageable hair. A simple chemical process explains the origin of hard water. Additional chemical reactions provide an explanation for the scum and deposits that readily form when soap and hard water combine.

THE CHEMICAL ESSENCE

More than 60% of the earth's water is groundwater, that is, water that travels through soil and rocks. Natural weathering of minerals such as limestone, calcite, shells, and coral is common as a consequence of the natural acidity of rainfall that results from dissolved atmospheric carbon dioxide. When this weathering process releases to the water supply positively charged metal ions that are constituents of minerals, the ions can combine with negatively charged ions present in soaps and detergents to form a waxy scum that does not dissolve in water.

The Chemical Specifics

Soap scum is an insoluble precipitate that forms between the cations of minerals typically present in hard water and the anions of soaps and detergents. Divalent cations of calcium (Ca^{2+}) and magnesium (Mg^{2+}) from calcium carbonate and magnesium carbonate minerals are the primary components of hard water. Divalent cations of iron (Fe^{2+}), manganese (Mn^{2+}), and strontium (Sr^{2+}) are also often present. An example of the dissolution (dissolving) process that releases calcium ions from calcium-containing minerals in contact with water with high acid levels is as follows:

$CaCO_3$ (s) (calcite, limestone, shells, coral) +
$$2 H^+ (aq) \rightarrow Ca^{2+} (aq) + H_2CO_3 (aq)$$

Soaps are composed of sodium salts of various fatty acids. These acids include those with the general structure $CH_3-(CH_2)_n-COOH$ where $n = 6$ (caprylic acid), 8 (capric acid), 10 (lauric acid), 12 (myristic acid), 14 (palmitic acid), and 16 (stearic acid). Oleic acid ($CH_3-(CH_2)_7-CH=CH-(CH_2)_7-COOH$) and linoleic acid ($CH_3-(CH_2)_4--CH=CH-CH_2-CH-COOH$) are also common soap ingredients. These sodium salts readily dissolve in water, but other metal ions such as Ca^{2+} and Mg^{2+} form precipitates with the fatty acid anions. For example, the dissolution of the sodium salt of lauric acid and the subsequent formation of a precipitate of the lauric acid anion with calcium ion is given by the following:

$$H_2O$$
$$CH_3-(CH_2)_{10}-COO^-Na^+ (s) \rightarrow CH_3-(CH_2)_{10}\text{-}COO^- (aq) + Na^+(aq)$$

$2 CH_3-(CH_2)_{10}-COO^- (aq) + Ca^{2+} (aq) \rightarrow$
$$(CH_3-(CH_2)_{10}-COO^-)_2 Ca^{2+} (s)$$

A variety of methods can be used to soften water. A cation exchange process is common in water softeners or conditioners. In these systems, hard water containing calcium and magnesium ions is passed through resin beads composed of styrene and divinylbenzene[1] and on whose surfaces are bound potassium and sodium ions. As the hard water passes through the resin beads, calcium and magnesium ions are retained on the resin surface

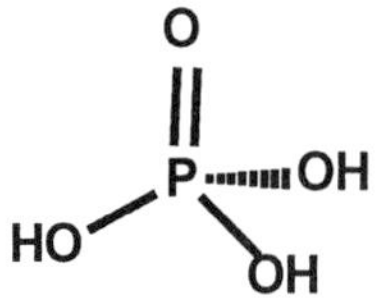

Figure 6   The chemical structure of phosphoric acid (H_3PO_4).

while sodium and potassium ions are released to the water. For charge balance, two sodium or potassium ions are released per divalent ion retained. As mentioned, the sodium and potassium cations do not form insoluble precipitates with the fatty acid anions in soap. To regenerate the beads, a strong NaCl or KCl saltwater solution is flushed through the softener. Calcium and magnesium ions are discharged as waste, and sodium and/or potassium ions are returned to the surface of the beads.

An alternative method of water softening involves adding chemical compounds known as *chelators* or *sequestrants*. Many household and institutional cleaners as well as personal care products contain glycolic acid ($HOCH_2COOH$)—also known as an alpha hydroxycarboxylic acid—to complex with metal cations to reduce the hardness of water.[2] The chelates formed from glycolic acid and cations such as calcium, magnesium, manganese, iron, and copper are water-soluble. Both the hydroxyl and carboxylic acid groups are used to form five-member ring complexes (chelates) with polyvalent metals. The formation of these chelated complexes frees the detergent from precipitating with the hard-water metal ions, thereby enhancing the detergent's ability to clean. Compounds classified as *polyphosphates* also serve as sequestrants in detergents. Derivatives of phosphoric acid (H_3PO_4) (Fig. 6) and pyrophosphoric acid ($H_4P_2O_7$) (Fig. 7) are common examples of sequestrants, specifically sodium acid pyrophosphate (disodium dihydrogen pyrophosphate, $Na_2H_2P_2O_7$), potassium (or sodium), acid phosphate (potassium or sodium dihydrogen phosphate, KH_2PO_4 or NaH_2PO_4), sodium hexametaphosphate (sodium polyphosphate, $(NaPO_3)_n \cdot Na_2O$), and tetrasodium pyrophosphate ($Na_4P_2O_7$).[3] A number of commer-

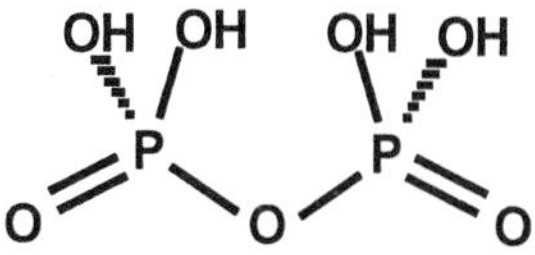

Figure 7   The chemical structure of pyrophosphoric acid ($H_4P_2O_7$).

cial products contain such sequestering agents, including Calgon dishwashing liquid, White Rain shampoo, and Spring Rain water and conditioner.

Unwanted calcium and magnesium ions can also be precipitated from water by adding washing soda ($Na_2CO_3 \cdot 10\,H_2O$), borax ($Na_2B_4O_7 \cdot 10\,H_2O$), or sodium silicate. Many commercial products contain these precipitating agents, including Arm and Hammer baking soda, Borateem laundry detergent, and Raindrops water conditioner.

KEY TERMS

dissolution; precipitation; chelator; sequestrant

REFERENCES

1. "What Makes Water Hard & How Hard Water Can Be Improved," Water Quality Association, http://www.wqa.org/Technical/Water-Hardness.html
2. "DuPont Specialty Chemicals: Glycolic Acid Applications," http://www.dupont.com/glycolicacid/applications/
3. "Markets & Applications: Water Treatment," Rhodia, Inc., http://www.rp.rpna.com/rhodia/searchmarket.icl

RELATED WEB SITES

"Chemical of the Week: Phosphoric Acid," University of Wisconsin, http://scifun.chem.wisc.edu/chemweek/h3po4/H3PO4.html
"Chemistry of Soaps and Detergents," Soap and Detergent Association, http://www.sdahq.org/
"Packaged Water Softeners," Rose Marie Tondl, http://www.ianr.unl.edu/pubs/NebFacts/NF94.htm
"Water Softening," Morton Salt, Morton International, http://www.mortonintl.com/salt/soft/sofasoft.htm

HOW DO HOUSEHOLD PRODUCTS SUCH AS DRANO AND LIQUID-PLUMR UNCLOG DRAINS?

Drains run slower and slower as greases, soaps, fats, and detergents build up on the inner walls of drain pipes and eventually cause blockage. What chemical action is used to unclog drains with household products such as Drano and Liquid-Plumr?

THE CHEMICAL ESSENCE

Most household drain cleaners are caustic substances that create *alkaline* (basic) solutions when dissolved in water. Certain organic compounds commonly called grease dissolve in water that is highly basic. Thus, the alkalinity (i.e., characterized by a pH value above 7) of drain cleaners is responsible for the dissolution of grease that has clogged drains and plumbing. Some crystal drain cleaners also contain solid aluminum particles that react with an alkaline solution to produce hydrogen gas and release heat. Both the agitation of evolving gas and the release of heat during the chemical reaction produce additional forces to open drains.

THE CHEMICAL SPECIFICS

Commercially available drain cleaners such as Drano and Liquid-Plumr contain sodium hydroxide (NaOH) as an active ingredient to dissolve grease.[1,2] Crystal drain cleaner contains the solid form of sodium hydroxide, whereas liquid drain cleaners are strong solutions of dissolved sodium hydroxide. In addition, some drain cleaners rely on the heat and vigorous bubbling produced from the action of strong base on finely divided particles of aluminum metal.[3-5] The overall reaction liberates hydrogen gas as follows:

Dissolution of strong base in water:
$$NaOH\ (s) \rightarrow Na^+(aq) + OH^-(aq)$$

Reaction of hydroxide ion with aluminum to form a soluble aluminum-containing ion and hydrogen gas:
$$2\ Al s) + 6\ H_2O(l) + 2\ OH^-(aq) \rightarrow 2\ Al(OH)_4^-(aq) + 3\ H_2(g)$$

The soluble ion $Al(OH)_4^-$ (aq) is known as the *aluminate* ion. The vigorous evolution of hydrogen gas helps to physically dislodge undissolved grease particles from the walls of plumbing.

KEY TERMS

dissolution; caustic; alkaline

REFERENCES

1. "Material Safety Data Sheet: Drano," Drackett Products Company, http://www.sanitary supplyco.com/m74610050.txt

2. "Material Safety Data Sheet: Liquid Drano," Drackett Products Company, http://siri.org/msds/h/q126/q115.html
3. "Material Safety Data Sheet: Liquid-Plumr," Clorox Company, http://siri.org/msds/h/q492/q211.html
4. Wood, J. T., and Eddy, R. M. (1996). What's the shiny stuff in Drāno? *J. Chem. Educ.* **73**, 463, http://jchemed.chem.wisc.edu/Journal/Issues/1996/May/abs463.html
5. "Reference Data Sheet for Chemical and Enzymatic Drain Cleaners," William D. Sheridan, http://www.mcs.net/~hutter/tee/draincle.html

WHAT IS SHATTERPROOF GLASS?

How does chemistry increase the resistance of glass to impact and shattering?

THE CHEMICAL ESSENCE

The shatterproof glass used in impact-resistant windows is actually not a glass material derived from silicon dioxide. Instead, shatterproof glass is a *thermoset plastic* or *thermoplastic* (i.e., a pliable material that is even easier to mold when hot). Shatterproof windows are composed of a specific thermoset material known as polycarbonate of bisphenol A (or bisphenol A polycarbonate). This clear, glassy polymer is constructed from repeating units of the monomer bisphenol A (Fig. 8). Thus, bisphenol A polycarbonate is also considered an example of a *heterochain polymer,* containing atoms in addition to carbon in its backbone chain. Bisphenol A polycarbonate is also a member of the larger polymer family of *polyesters*, that is, polymers formed from a large number of smaller molecules, or monomers, by establishment of ester linkages (Fig. 9) between them. General Electric markets the polycarbonate of bisphenol A as Lexan.[1] Similar bisphenol A polycarbonate sheets are marketed by Rohm and Haas as the product Tuffak.[2]

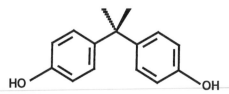

Figure 8   The repeating units of the monomer bisphenol A used to construct the polymeric thermoset material used in shatterproof glass.

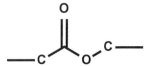

Figure 9   The ester linkage found in the polymers known as polyesters.

THE CHEMICAL SPECIFICS

Polycarbonate is a generic term for the class of polymers consisting of long-chain linear polyesters of carbonic acid, H_2CO_3, and aromatic alcohols known as phenols that possess two hydroxyl groups.

The synthesis of polycarbonate of bisphenol A begins with the reaction of bisphenol A and sodium hydroxide to obtain the sodium salt of bisphenol A, as in Figure 10. The sodium salt of bisphenol A is then reacted with phosgene to produce the polycarbonate, as diagrammed in Figure 11.[3] Alternative synthesis procedures under exploration replace the solvent phosgene with carbon dioxide to accomplish the final step of the synthesis.[4]

The properties of the polycarbonate of bisphenol A are directly related to the structure of the polymer. The molecular stiffness associated with this polycarbonate arises from the presence of the rigid phenyl groups on the molecular chain or backbone of the polymer and the additional presence of two methyl side groups. The transparency of the material arises from the amorphous (noncrystalline) nature of the polymer. A significant crystalline structure is not observed in the polycarbonate of bisphenol A because intermolecular attractions between phenyl groups of neighboring polymer chains lead to a lack of mobility of the chains that deters the development of a crystalline structure.

Another polymer used for unbreakable windows is poly(methyl methacrylate). PMMA is a vinyl polymer, made by free radical vinyl polymerization from the monomer methyl methacrylate, according to the reaction in Figure 12.[3] Rohm and Haas markets this PMMA-based shatterproof glass as Plexiglas.[5] Imperial Chemical Industries also produces PMMA as Lucite.

KEY TERMS

polycarbonate; polyester; polymer; monomer; thermoset plastic; thermoplastic; heterochain polymer

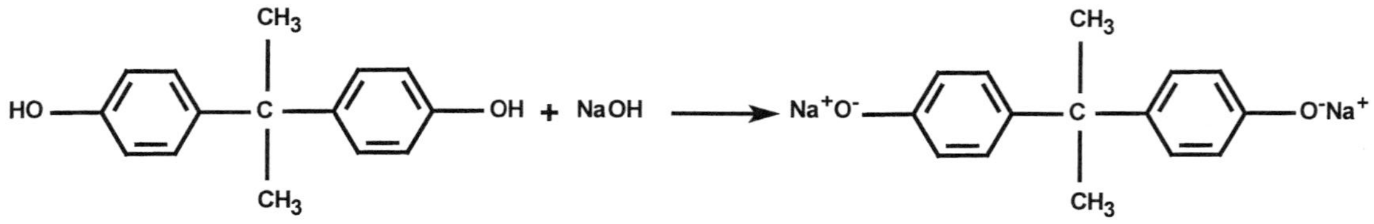

Figure 10 The reaction of bisphenol A and sodium hydroxide to obtain the sodium salt of bisphenol A.

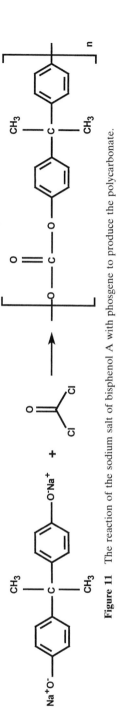

Figure 11   The reaction of the sodium salt of bisphenol A with phosgene to produce the polycarbonate.

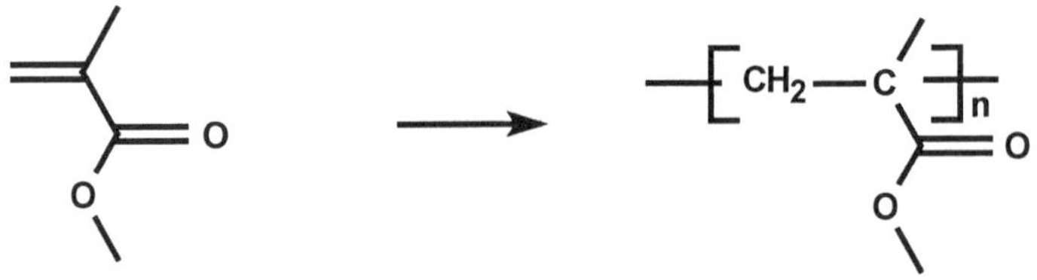

Figure 12 Free radical vinyl polymerization of the monomer methyl methacrylate to form the vinyl polymer poly(methyl methacrylate) (PMMA).

REFERENCES

1. "Material Safety Data Sheet, Lexan Resin," General Electric, http://www.centor.com/cb-matcd/plast4.html
2. "Material Safety Data Sheet, TUFFAK Polycarbonate Sheet," Rohm and Haas, http://www.rohmhaas.com/atohaas/msds/a900342.htm
3. "Polycarbonates," Department of Polymer Science, University of Southern Mississippi, http://www.psrc.usm.edu/macrog/pc.html
4. "Production of Bisphenol-A Polycarbonate Using Carbon Dioxide," Pacific Northwest Pollution Prevention Research Center, http://pprc.pnl.gov/pprc/rpd/fedfund/doe/doe_oit/product.html
5. "Material Safety Data Sheet Plexiglas(R) Impact Grade Acrylic Resin," Rohm and Haas, http://www.rohmhaas.com/atohaas/msds/a872406.htm

RELATED WEB SITES

"Making Polycarbonate," Department of Polymer Science, University of Southern Mississippi, http://www.psrc.usm.edu/macrog/pcsyn.html

HOW DO WINDSHIELD COATINGS IMPROVE VISIBILITY IN A RAINSTORM?

> "Disperses rain on contact, keeping windshield clear and clean. See better. Drive safer. With or without wipers."[1]

How can a product claim to improve your vision through an automobile windshield? The chemical structure of the key ingredients in the formulation can ensure a clear view during a rainstorm.

THE CHEMICAL ESSENCE

A number of glass treatments are available for both automobiles and aircraft to disperse rain, sleet, and snow; prevent the buildup of frost, ice, salt, mud, bugs, oil, and road grime; and reduce glare. These glass coatings are durable but not permanent and can be reapplied. These rain-repellent materials contain *hydrophobic* substances that interact strongly with glass but readily shed rain drops.

THE CHEMICAL SPECIFICS

The manufacturers of windshield coatings take advantage of the principle that *hydrophilic* substances possess a chemical structure that permits favorable intermolecular interactions with water. Chemical species capable of exhibiting hydrogen bonding, dipole-dipole interactions, or ion-dipole interactions with water are typically hydrophilic substances. Alternatively, *hydrophobic* substances typically are nonpolar molecules that exhibit only weak van der Waals interactions with water.

Two major classes of hydrophobic chemical substances can be applied to glass in ultrathin layers to inhibit surface wetting. *Siloxanes* or *polysiloxanes* or *silicones* are polymers with a "backbone" of alternating silicon and oxygen atoms. These macromolecules are quite chemically inert, show resistance to water, and exhibit stability at high and low temperatures. The most common siloxane polymer, polydimethylsiloxane, is composed of the monomeric (i.e., repeating) unit illustrated in Figure 13. A hydroxy terminated poly(dimethylsiloxane), $HO[-Si(CH_3)_2O-]_nH$, is one of the constituents of the formulation known as Rain-X.[2]

The halogenated hydrocarbons known as *chlorofluorocarbons* are a second class of materials that comprise water-repellent coatings. In particular, the substance known as 1,1,2-trichloro-1,2,2-trifluoroethane or CFC-113

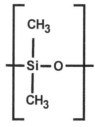

Figure 13 The monomeric (i.e., repeating) unit of the most common siloxane polymer, polydimethylsiloxane.

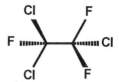

Figure 14   1,1,2-trichloro-1,2,2-trifluoroethane or CFC-113

(Figure 14) is the main ingredient of one windshield product.[3] Other proprietary fluorinated compounds are used in patented technology to produce a highly effective windshield glass treatment.[4]

KEY TERMS

hydrophobic; hydrophilic; siloxanes; chlorofluorocarbons

REFERENCES

1. Auto-Gard Rain Repellent, Auto-gard Premium Car Care, http://www.miles-levels.com/autogard.html
2. "Unelko Corporation: Rain-X Material Safety Data Sheet," http://www.unelko.com/rxmsds.html
3. "Potential Rainboe Toxicity in Commercial Aircraft Cockpits," Victoria Voge, MD, MPH, Gonzales, Tex; and Lance Schaeffer, JD, http://www.sma.org/smj/96aboccp.htm
4. "Aquapel Glass Treatment News: Research Highlights Improved Safety Potential of Hydrophobic Windshield Coatings," PPG Industries, Inc., http://www.aquapel.com/

RELATED WEB SITES

"Windshield coatings do keep the rain away," http://projo.com/wheels/maintain/1019ma1.htm

WHAT CAUSES THE PEARLESCENT APPEARANCES OF SOME PAINTS?

Pearlescent lusters are seen in many paints, inks, and cosmetics. The pearlescent pigment technology that brings us these unusual effects relies on a common mineral to achieve these opalescent qualities.

THE CHEMICAL ESSENCE

Pearlescent pigments contain small flakes or platelets of the mineral mica that are additionally coated with a very thin layer of titanium dioxide. The simultaneous reflection of light from many layers of small platelets creates an impression of luster and sheen. By varying the thickness of the coating on the surface of the mica particles, pigment manufacturers can achieve a range of colors for the pearlescent effect.

THE CHEMICAL SPECIFICS

Pearlescent pigments are derived from microscopic mica platelets ranging in size from 1 to 2 microns in thickness and up to 180 microns in diameter. Mica consists of an aluminum silicate with a crystalline structure that easily permits cleavage into very thin platelets. In fact, mica is a generic term for any one of several complex hydrous aluminosilicate minerals characterized by their platy nature and pronounced basal cleavage. The general formula for mica is AB_{2-3} (Al, Si) Si_3 O_{10} (F, OH)$_2$.[1] Generally A represents the metal potassium (K) and B corresponds to aluminum (Al). But some micas contain calcium (Ca), sodium (Na), or barium (Ba) for A and lithium (Li), iron (Fe) , or magnesium (Mg) for B. This subclass of silicates contains rings of SiO_4 tetrahedrons linked by shared oxygens to other rings in a two-dimensional plane. This bonding arrangement produces a sheetlike structure that gives rise to flat, platy crystals with good basal cleavage.[1]

The presence of mica in pearlescent pigments only partly accounts for the appearance of the pigment. A very thin layer of the inorganic oxide titanium dioxide (TiO_2) or iron oxide (Fe_2O_3) or both is coated on the mica platelets. The various colors and pearlescent effects are created as light is both refracted and reflected from the titanium dioxide layers. The very thin platelets are highly reflective and transparent. With their platelike shape, the platelets are easily oriented into parallel layers as the paint medium is applied. Some of the incident light is reflected from the uppermost layers, while a portion is transmitted to lower layers and then reflected. A pearlescent luster is produced from this multiple reflection of light from many microscopic layers. Smaller mica particle size leads to smoother sheens, whereas larger particle size produces a higher luster or sparkled effect. In addition, the thickness of the oxide layer dictates the color observed. Multiple reflections and refractions of light lead to both constructive and destructive interference of light waves. The oxide layer thickness determines the narrow range of wavelengths of light that interfere constructively,

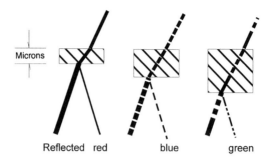

Microns

Reflected red   blue   green

Figure 15 The multiple reflection of light from microscopic oxide layers of different dimensions leads to constructive and destructive interference of light waves, producing a particular color effect. Different thicknesses reflect different colors.

thus producing a particular color effect. All other wavelengths of light interfere destructively and are not observed. Figure 15 summarizes these concepts.

KEY TERMS

interference; inorganic oxide

REFERENCES

1. "The Silicate Class," The Mineral Gallery, http://www.galleries.com/minerals/silicate/class.htm

RELATED WEB SITES

"Interference and Iridescent Acrylics," Golden Artist Colors, New Berlin, NY, http://www.goldenpaints.com/iridint.htm

Microscopic photo of pearlescent paint, http://www.standox.de/standote/englisch/color/3schicht.htm

"Standothek: Design and Paint," http://www.standox.com/standote/englisch/design/effekt.htm

"Understanding Vehicle Paint Technology," http://www.standox.de/lack/englisch/guide3.htm #Metallic

WHY ARE THE SMALL PACKS OF GRANULES LABELED "DESICCANTS" ENCLOSED IN BOTTLES OF MEDICATION, NEW SHOES, AND ELECTRONICS? WHAT IS THE IDENTITY OF THE DESICCANTS?

Have you ever wondered about the content of the small packets of granules included in boxes of new electronic devices, leather products, or medications? Or have you simply heeded the warning label to dispose of the sachet immediately? The granular desiccants in these packets have particular physical properties that enhance their function as drying agents. A look at the chemical structure or chemical properties of these materials provides a better understanding of their capacity to control moisture.

THE CHEMICAL ESSENCE

Perhaps you have just purchased a new electronic device, such as a personal computer, video camera, pager, or cell phone. The advanced circuitry in such devices can malfunction in humid environments. Moisture can reroute and even shortcircuit electric signals, impairing the operation of your new electronic equipment. Manufacturers recognize that moisture can adversely affect their products and therefore include small packets of *desiccant* to control the moisture levels during shipping and storage. A desiccant is a porous solid drying agent or hydrating agent that attracts moisture from the atmosphere and retains those particles of water on its surface or in the pores of the desiccant. Both synthetic materials and naturally occurring minerals function as desiccants, including dehydrated gypsum, calcinated lime, and a type of clay. In addition to electronic devices, a myriad of other products benefit from the application of desiccants, including food, pharmaceuticals, shoes and other leather articles, laboratory equipment and hygroscopic chemicals, books and rare manuscripts, museum and historical artifacts, paintings and valuable art objects, film, hearing aids, and stamps.

THE CHEMICAL SPECIFICS

Five common desiccant materials are used to adsorb water vapor: montmorillonite clay ($(Na,Ca_{0.5})_{0.33}(Al,Mg)_2Si_4O_{10} \cdot nH_2O$), silica gel, molecular sieves (synthetic zeolite), calcium sulfate ($CaSO_4$), and calcium oxide (CaO). These desiccants remove water by a variety of physical and chemical methods: *adsorption,* a process whereby a layer or layers of water molecules adhere to the surface of the desiccant; *capillary condensation,* a procedure whereby the small pores of the desiccant become filled with water; and *chemical action,* a procedure whereby the desiccant undergoes a chemical reaction with water.

Montmorillonite clay is a naturally occurring adsorbent that swells to several times its original volume when water adsorption occurs.[1] The most commonly used desiccant is silica gel ($SiO_2 \cdot H_2O$), an amorphous form of silica manufactured from sodium silicate and sulfuric acid. The porous nature of silica gel forms a vast surface area that attracts and holds water by both adsorption and capillary condensation, allowing silica gel to adsorb about 40% of its weight in water.[2] Zeolites or "molecular sieves" are rigid, hydrated crystalline aluminosilicate minerals that contain alkali and alkaline earth metals. Zeolites possess a three-dimensional crystal lattice structure that forms surface pores of uniform diameter and contain numerous regular internal cavities and channels. Water molecules are readily incorporated within the pores and cavities. The zeolite structure consists of interlocking tetrahedrons of $[SiO_4]^{4-}$ and $[AlO_4]^{5-}$. Each oxygen atom is shared by two tetrahedra. For charge balance, other metal ions are present in the cavities of the framework, typically monovalent or divalent ions such as sodium, potassium, magnesium, calcium, and barium. The drying effect of calcium sulfate (dehydrated gypsum) occurs by chemical action, as the hemihydrate of calcium sulfate is created when the anhydrous calcium sulfate reacts with water.

$$2\ CaSO_4 + H_2O \rightarrow 2\ CaSO_4 \cdot \tfrac{1}{2}\ H_2O$$

Calcium chloride (also known as calcinated lime) acts as a drying agent as a consequence of its *deliquescent* nature—solid $CaCl_2$ readily absorbs water from the atmosphere and subsequently dissolves.

KEY TERMS

desiccant; anhydrous; hydrate; deliquescent

REFERENCES

1. The Mineral MONTMORILLONITE, Amethyst Galleries, Inc., http://mineral.galleries. com/minerals/silicate/montmori/montmori.htm
2. "Selecting the Right Desiccant," http://www.multisorb.com/faqs/selecting.html

RELATED WEB SITES

"Components of printing inks," http://www.fhd-stuttgart.de/projekte/printing-inks/p_ compo0.htm#inorganicpigments

"Frequently Asked Questions," Desiccare, Inc., http://206.190.82.22/faq.htm

"Protection of Pharmaceuticals and Diagnostic Products through Desiccant Technology," Rodney L. Dobson, Multisorb Technologies, Inc., http://www.multisorb.com/faqs/ protection.html

"Prudent Food Storage: Questions & Answers:Desiccants," http://chetday.com/fsfaq8. html#d&d1

"The Zeolite Group," Amethyst Galleries, Inc., http://galleries.com/minerals/silicate/ zeolites.htm

"Zeolite: The Versatile Mineral," ZeoponiX, Inc., http://www.zeoponix.com/html/body_ zeolites.html

WHAT ARE THE RED OR SILVER LIQUIDS IN THERMOMETERS?

We're accustomed to the rise and fall of the liquid in a thermometer as temperatures increase and decrease, respectively. Why do these liquids respond to temperature in this fashion, and what are the identities of these materials?

THE CHEMICAL ESSENCE

The invention of the thermometer is generally credited to Galileo. His instrument, built near the end of the 16th century, relied on the expansion of air with an increase of heat. Traditional liquid-in-glass thermometers were devised in the 1630s and are standard equipment today in research settings, medical practice, and meteorological measurement.

Many common thermometers contain a liquid confined within a narrow capillary tube. The liquid height varies with the surrounding temperature. In actuality, the *volume* of the liquid is responding to temperature, and the liquid tries to expand equally in all directions. By confining the liquid in a tube, the only direction for ready expansion is along the length of the narrow tube. Thus, expansion in that direction (i.e., liquid height) can be used as a measure of the ambient temperature. Most liquids expand in volume as their temperature increases, and, because the extent of expansion is generally constant over a range of temperatures, the amount of expansion can be quantified and calibrated.

In particular, two liquids exhibit a consistent and measurable expansion at commonly measured temperatures—liquid mercury and ethanol (also known as ethyl alcohol). Daniel Gabriel Fahrenheit (1686–1736), a German physicist and maker of scientific instruments, is credited for inventing the alcohol thermometer in 1709 and the mercury thermometer in 1714 (as well as developing the temperature scale that bears his name). These liquids permit common temperatures to be readily measured, such as the boiling and freezing of water. Why are these particular measurements possible? Mercury has a *higher* boiling point than water, and ethanol has a *lower* freezing point than water. The silvery color of mercury facilitates viewing the liquid level in a thermometer, but the colorless appearance of ethanol generally is modified with a red dye to enhance distinguishing the liquid level.

THE CHEMICAL SPECIFICS

The useful temperatures ranges for mercury and ethanol-filled thermometers depend on the temperature ranges over which these materials remain as liquids. Mercury exhibits freezing and boiling points at atmospheric pressure (i.e., *normal* freezing and boiling points at 1 atm) of $-38.9°C$ and $356.6°C$. Thus, a mercury thermometer is advantageous when high temperature conditions are likely. Ethanol, with normal freezing and boiling points of $-114.1°C$ and $78.3°C$, respectively, is convenient as a thermometer liquid when temperatures below the freezing point of water are to be measured.

The equation relating the volume change of a material to a change in temperature is as follows:

$$\Delta V = \beta V_o(\Delta T) \text{ or } V_f = V_o (1 + \beta \Delta T)$$

for V_o at an initial T of $0°C$ and where β is the volume expansion coefficient or the coefficient of cubic expansion or the coefficient of thermal expansion. The units of β are reciprocal temperature, for β is defined as the increase in volume per unit volume per degree C rise in temperature. Mercury has an average value for β of $1.8169041 \times 10^{-4}°C^{-1}$ over the temperature range of 0 to $100°C$ and $1.81163 \times 10^{-4}°C^{-1}$ over the temperature range of 24 to $299°C$.[1] The volume expansion coefficient β for ethanol averages $1.04139 \times 10^{-3}°C^{-1}$ over the temperature range of 0 to $80°C$ [1], a larger value allowing for finer calibrated thermometers. An important point to note is that any expansion or contraction of the thermometer container itself (usually glass) is generally ignored when calibrating household thermometers because liquids generally have a substantially larger coefficient of thermal expansion than do solids. As an example, Corning 790 glass exhibits a cubical expansion coefficient of $2.4 \times 10^{-6}°C^{-1}$.[2] Pyrex glass contains borosilicate glass, a type of glass that is exceptionally resistant to heat, expanding only about one-third as much as common silicate glass. As a consequence, Pyrex is often used to make chemical apparatus, including thermometers.

KEY TERMS

normal boiling point; normal freezing point; volume expansion coefficient (or coefficient of cubic expansion)

REFERENCES

1. Dean, J. A. (Ed.). (1979). Coefficients of cubical expansion for various liquids and aqueous solutions. *In: Lange's handbook of chemistry,* 12th ed., Table 10-42. McGraw-Hill, New York.
2. Dean, J. A. (Ed.). (1979). Coefficients of cubical expansion of solids. *In: Lange's handbook of chemistry,* 12th ed., Table 10-43. McGraw-Hill, New York.

RELATED WEB SITES

"The Thermometer," Albert Van Helden, http://es.rice.edu/ES/humsoc/Galileo/Things/thermometer.html

WHY DOES A KITCHEN GAS BURNER GLOW YELLOW WHEN A POT OF BOILING WATER OVERFLOWS?

A cook recognizes the telltale signs of an overflowing pot of boiling water—the characteristic hiss as the water hits the hot gas burner and evaporates and the familiar accompanying yellow glow. Chemistry on the atomic level is responsible for the brilliant yellow illumination.

THE CHEMICAL ESSENCE

The yellow color imparted to a natural gas flame originates from the ignition of sodium atoms or ions. The common source of sodium is salt (sodium chloride) naturally dissolved in water or at higher concentrations from the food being heated. (Have you ever noticed that the overflowing water that evaporates on the hot gas burners often leaves a white residue? That white solid is dried sodium chloride salt.) Hot sodium ions emit light of characteristic colors or frequencies, an unambiguous indicator of the presence of this element.

THE CHEMICAL SPECIFICS

The element sodium is a member of the alkali metal family. Thermal ionization of alkali metals is possible in very hot flames because of the low ionization potentials of these metals. Atomic sodium, for example, has an ionization energy of 495.8 kJ mol^{-1} ($\approx$5 electron volts).[1] Recall that the ionization energy of a neutral atom is defined as the energy required to remove the lowest-energy electron from the gaseous atom and form the positive cation: Na (g) $\rightarrow$ Na$^+$ (g) + e$^-$. The sodium ion and electron may then recombine to form a neutral but excited sodium atom. Following this thermal excitation of sodium atoms to high energy states, sodium returns to the ground state via the emission of photons. The energy of these photons is equivalent to the wavelength of light observed. The most intense emission occurs in the yellow region of the visible range of the electromagnetic spectrum at wavelengths of 589.0 and 596.6 nm.[2]

We observe the typical yellow emission of sodium atoms in many other circumstances. Sodium vapor lamps are electric discharge lamps with metal electrodes and filled with neon gas and a small amount of sodium. Current passing through the electrodes first ionizes the neon gas. The hot neon gas then vaporizes the sodium, which is then easily excited. The characteristic yellow light emanates from the excited state species returning to the ground state. Astronomers recently discovered that one of the three tails of particles streaming from the Comet Hale-Bopp was a bright yellow tail of emitting sodium atoms.[3] Scientists at the Jet Propulsion Laboratory in Pasadena, California, also reported a yellowish emission from Jupiter's moon Io arising from the cloud of sodium vapor forming a halo around Io.[4]

KEY TERMS

atomic emission spectra; ionization energy; alkali metals

REFERENCES

1. "Sodium," http://wild-turkey.mit.edu/Chemicool/elements/sodium.html
2. Lide, D. R. (Ed.). (1993). Line spectra of the elements. In: *Handbook of chemistry and physics*, 74th ed., p. 10-92. CRC Press, Boca Raton, FL.

3. Three-tailed comet. (1997, May 3). *New Scientist Planet Science in Brief,* http://www.nsplus. com/ns/970503/inbrief.html

4. Photo caption, Jet Propulsion Laboratory, California Institute of Technology, National Aeronautics and Space Administration, Pasadena, CA, http://nssdc.gsfc.nasa.gov/photo_ gallery/caption/gal_io3_48584.txt

RELATED WEB SITES

"Chemical of the Week: Gases That Emit Light," http://scifun.chem.wisc.edu/CHEMWEEK/ gasemit/gas-emit.html

WHY DOES SUPERGLUE STICK TO ALMOST EVERY SURFACE?

One commercial adhesive is marketed with the following claims:

"*High Strength Adhesive *Durable Bonding *Fast Acting *Bonds Metals, Rubber, Ceramics, Plastics, Glass, Wood, Veneers, Fabrics, Vinyl, Cardboard, Cork, Leather, Nylon, and Other Similar Surfaces"[1]

How can one substance act as a general purpose adhesive with affinity for so many types of surfaces?

THE CHEMICAL ESSENCE

One class of adhesives known as *superglues* consists of synthetic organic polymers that provide strong and rapid adhesion. These adhesives are unusual in that the polymerization process to form the adhesive occurs upon exposure of the monomer to water. Under most conditions, atmospheric moisture is sufficient to form a strong adhesive. Superglues stick to a variety of surfaces because a film of moisture exists on almost any surface. The quality of bonding varies with the humidity; the higher the humidity, the better the set.

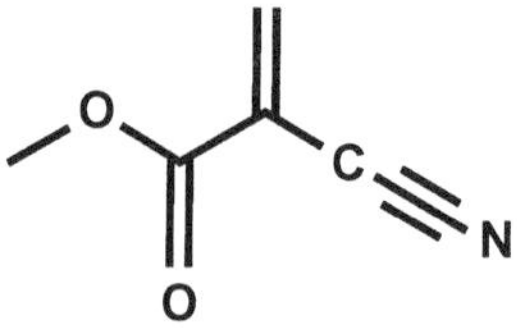

Figure 16   The monomer methyl α-cyanoacrylate found in the adhesive marketed under the trademark "Superglue."

THE CHEMICAL SPECIFICS

The adhesive marketed under the tradename Superglue contains the monomer methyl α-cyanoacrylate (Fig. 16). A variety of cyanoacrylates are commercially sold as contact adhesives with the alkyl group −R denoted in Figure 17 varying from a methyl group to produce ethyl, isopropyl, allyl, butyl, isobutyl, methoxyethyl, and ethoxyethyl cyanoacrylate esters.[1-4] The properties of the adhesive (e.g., setting time, strength, durability) vary with the substitution. All of these monomers undergo polymerization in the presence of water. In fact, water serves as an *initiator* for the polymerization according to an anionic vinyl polymerization mechanism.[5] The overall polymerization process to form the cyanoacrylate polymer is represented schematically by the reaction in Figure 18.

Although the formation and subsequent cross-linking of the polymer is one factor in the effectiveness of an adhesive, the adhesive strength of the polymer-surface interface is also critical. Both physical and chemical considerations influence the bonding. A rough or porous surface is generally more effective for "locking" an adhesive to a substrate. But "chemically active" sites on the surface to promote interaction via hydrogen bonding, strong dipole-dipole interactions, or other intermolecular attractions also contribute to the adhesion properties.

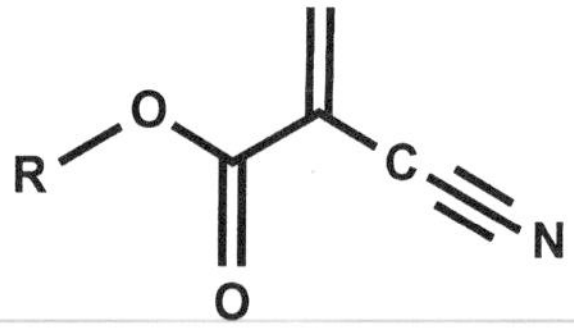

Figure 17   The general class of cyanoacrylates with varying alkyl group -R, sold as contact adhesives.

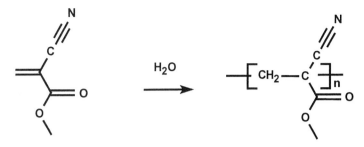

Figure 18 The overall polymerization process to form the cyanoacrylate polymer.

KEY TERMS

monomer; polymer; adhesion

REFERENCES

1. "Tech Hold© Cyanoacrylate," http://www.techspray.com/2502.htm
2. "Polycyanoacrylates," Department of Polymer Science, University of Southern Mississippi, http://www.psrc.usm.edu/macrog/pca.html
3. Technical data sheet for Gel Set(c) 44 Ethyl Cyanoacrylate,http://www.holdtite.com/english/technical/tech/ca44.htm
4. "DynaBond Adhesive," http://www.dynamesh.com/dba600msds.htm
5. "Anionic Vinyl Polymerization," Department of Polymer Science, University of Southern Mississippi, http://www.psrc.usm.edu/macrog/anionic.html

RELATED WEB SITES

"Reactive adhesives, proactive chemistry," http://www.chemsoc.org/gateway/chembyte/cib/glue.htm
"Superglue," Bruce Sterling, http://www.eff.org/pub/Publications/Bruce_Sterling/FSF_columns/fsf.07

WHY ARE ICE CUBES CLOUDY ON THE INSIDE?

Pure, clear liquid water should freeze as a clear solid, but ice cubes commonly have a cloudy appearance. What chemistry should you know to recognize that cloudy ice cubes are not cause for serious alarm?

THE CHEMICAL ESSENCE

Water is rarely a "pure" substance but contains both dissolved gases (e.g., oxygen) from the atmosphere and dissolved minerals (e.g., calcium and magnesium salts). The presence of these substances affects the temperature at which water freezes. Pure water freezes at 0°C; water with dissolved gases and mineral salts freezes at a lower temperature. The higher the concentration of dissolved gases and minerals, the lower the freezing point of water. As water cools, the first layer of ice that forms is at the interface with air. As ice forms, *pure water* solidifies, leaving the dissolved gases and salts in solution. Thus, the freezing process concentrates the dissolved species in smaller and smaller volumes of liquid solution, effectively increasing their concentration . With a higher concentration of dissolved material, the temperature at which additional ice will form is lowered. The cloudiness in the center of an ice cube thus is the consequence of the concentration of dissolved gases and minerals that refract light and create an opaque appearance.

THE CHEMICAL SPECIFICS

The depression of the freezing point of a solvent due to the presence of a dissolved solute is an example of a *colligative property,* that is a property of a dilute solution that depends on the number of dissolved particles and not on the identity of the particles. Water has a *freezing point depression constant,* K_f, of 1.86 K kg mol^{-1}. In other words, for every mole of nonvolatile solute dissolved in a kilogram of water, the freezing point of water is lowered by 1.86°C. The change in freezing point, ΔT, can be calculated from the equation:

$$\Delta T = K_f \, m$$

where m is the molality of the solution, the number of moles of solute dissolved per kilogram of solvent (water). If the solute dissociates into ions upon dissolution in water, then m must be expressed as the *total molality* of all species (nondissociating or ionic) in the solution. Thus, one mole of NaCl in a kilogram of water will lower the freezing point by 2×1.86°C as a consequence of the two moles of ions (Na^+ and Cl^-) present in solution. In contrast, one mole of sugar in one kilogram of water would lower the freezing point by 1.86°C, for sugar does not dissociate into ions when dissolved.

KEY TERMS

freezing point depression; molality; colligative property; solvent; solute

RELATED WEB SITES

"Colligative Properties," Ryan Pearman, Department of Chemistry, University of Illinois, Urbana-Champaign, http://dionysus.phs.uiuc.edu/~pearman/101Online/notes/Colligative/Colligative.html

"On the Theory of Electrolytes. I. Freezing Point Depression and Related Phenomena. (Zur Theorie der Elektrolyte. I. Gefrierpunktserniedrigung und verwandte Erscheinungen)," P. Debye and E. Hückel, translated from *Physikalische Zeitschrift,* Vol. 24, No. 9, 1923, pp. 185–206, Classic Papers from the History of Chemistry, http://dbhs.wvusd.k12.ca.us/Chem-History/Debye-Strong-Electrolyte.html

Chapter 8

Connections to the Theater and the Arts

WHAT IS THE ORIGIN OF THE EXPRESSION "IN THE LIMELIGHT"?

We often use the expression "in the limelight" to characterize someone at the center of attention. Theatrical production in the mid-19th century created "limelight" through a chemical reaction to illuminate the stage and assist theatergoers in viewing the star performers.

THE CHEMICAL ESSENCE

Limelights were a lighting system invented by the British engineer Captain Thomas Drummond in 1816 to use for surveying purposes. These novel lighting systems were essentially very bright gas lamps that used the heated element rather than the gas flame to generate light. When heated in a flame consisting of jets of oxygen and hydrogen gas (an oxyhydrogen flame), a block of *lime* (calcium oxide) becomes incandescent and emits a soft, brilliant white light. (*Incandescence* is a general term to describe light produced by heating a solid. Incandescent lightbulbs heat solid tungsten filaments to produce light.) While the sharp point of an oxyhydrogen flame generates a small area of incandescence, mirrored reflectors can be used to direct the intense light and expand the area of illumination. Limelights were first used in the theater in 1837 and were widely employed by the 1860s.[1] The visibility of the light over long distances (more than 66 miles according to Drummond)[2] led to the use of limelight in lighthouses. For example, in 1861 the illuminating properties of the limelight were tested in the lighthouse on

135

the white chalk cliffs of the South Foreland, the closest point of approach of mainland England to France.[3] The ability to achieve remote lighting on stage with a powerful source that could be focused and varied in brightness for special effects largely contributed to the popularity of this lighting system. Illumination of front and center stage by limelights in the form of "spotlightling" led to the reference of the most desirable acting area on stage as "in the limelight." As the lime is consumed by burning, continual illumination requires an operator to constantly supply the flame with a fresh surface of lime, a drawback of this lighting method.

THE CHEMICAL SPECIFICS

Calcium oxide can be produced from extensive heating of limestone. Primarily composed of calcium carbonate, limestone is extracted from both underground and surface mines and heated to temperatures exceeding 180°F to convert the calcium carbonate into calcium oxide. This *thermal decomposition* reaction also generates carbon dioxide gas:

$$CaCO_3 \text{ (s)} \rightarrow CaO \text{ (s)} + CO_2 \text{ (g)}$$

Calcium oxide crystallizes in the sodium chloride structure, a structure with two interpenetrating *face-centered cubic lattices*. In the NaCl lattice depicted in Figure 1, the face-centered cubic arrangement of sodium cations (the smaller spheres) is readily apparent with the larger spheres (representing chloride anions) filling what are known as the octahedral holes of the lattice. Octahedral holes are defined as cavities in a crystal lattice that have six identical and equidistant atoms or ions as the nearest neighboring species. The NaCl structure is also characterized by an arrangement that allows each cation to have six equidistant anions as nearest neighboring ions and each anion to have six equidistant neighboring cations. Thus, both the sodium and chloride ions are said to have a *coordination number* of six.

The chemistry of calcium oxide limits the lifetime of the limelight. Exposure at ordinary temperatures to water moisture and carbon dioxide in the atmosphere ultimately yields calcium carbonate, which fractures upon strong heating.

$$CaO \text{ (s)} + H_2O \rightarrow Ca(OH)_2 \text{ (s)}$$
$$Ca(OH)_2 \text{ (s)} + CO_2 \text{ (g)} \rightarrow CaCO_3 \text{ (s)} + H_2O \text{ (l)}$$
$$CaO \text{ (s)} + CO_2 \text{ (g)} \rightarrow CaCO_3 \text{ (s)}$$

It was common practice in the 19th century to wrap the lime block in dense paper or with a coating of wax to prolong the use of the lighting element.[3] Storage of the lime in a sealed can to limit contact with air also extended its utility.

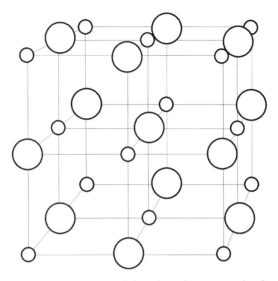

Figure 1   The NaCl crystal structure consisting of two interpenetrating face-centered cubic lattices. The face-centered cubic arrangement of sodium cations (the smaller spheres) is readily apparent with the larger spheres (representing chloride anions) filling what are known as the octahedral holes in the lattice. Calcium oxide also crystallizes in the sodium chloride structure.

KEY TERMS

lime; face-centered cubic crystal lattice; coordination number; octahedral hole; thermal decomposition

REFERENCES

1. "Limelight," *Britannica Online,* http://www.eb.com:180/cgi-bin/g?DocF=micro/349/92.html
2. Hocking, M. B., and Lambert, M. L. (1987). A reacquaintance with the limelight. *J. of Chemical Education* **64,** 306–310.
3. "World Lighthouse Information: The Lighthouses of England: South Foreland Contents: Experiments in Illumination," http://www.btinternet.com/~k.trethewey/

RELATED WEB SITES

"Light   Producing   Technologies,"   http://www.public.iastate.edu/~rprusk/htmls/LightTechnologies.html
"Lime: The Essential Chemical," The National Lime Association, http://www.lime.org/aboutlime.html

WHY DOES A FLASHBULB DEVELOP A WHITE COATING AFTER A FLASH?

"A . . . brilliant light . . . can be obtained by burning . . . magnesium in oxygen. A piece of magnesium wire held by one end in the hand, may be lighted at the other extremity by holding it to a candle. . . . It then burns away of its own accord evolving a light insupportably brilliant to the unprotected eye."

—*William Crookes, editor of the* Photographic News, *October 1859*[1]

THE CHEMICAL ESSENCE

Flashbulbs and flash lamps are devised to produce intensely brilliant but brief emissions of light for a variety of applications. Photographic flashbulbs contain a mesh of fine magnesium wire in an oxygen-rich atmosphere. When the flashbulb is activated, an electric current passes through the wire, heating the metal, triggering a reaction with oxygen to produce a white, powdery solid known as magnesium oxide that coats the interior of the bulb.

THE CHEMICAL SPECIFICS

The first portrait using magnesium was taken by Alfred Brothers of Manchester on February 22, 1864.[1] However, a limited understanding of the chemistry of magnesium and its prohibitive cost inhibited the development of this application. Advances by Robert Bunsen (the discoverer of the element cesium and for whom the Bunsen burner is named) led to more economic production of pure magnesium by electrolytic means.[2] Bunsen also advocated the use of magnesium for illumination purposes. In the late 1880s the ignition of magnesium powder with an oxidizing agent such as potassium chlorate was discovered and led to the introduction of flash powder. Professional photographers then produced brilliant flashes of light (as well as acrid smoke and ash) by firing finely powdered magnesium with a percussion cap.

The development of photographic flashbulbs occurred in the 1920s. These devices generally consist of a transparent glass bulb filled with a mesh of aluminum, magnesium, or zirconium metal in a volume of oxygen gas. When these metals burn in air (i.e., undergo combustion with oxygen), they produce an intense white light—the flash we observe in a flashbulb. But the combustion of these metals with oxygen is a slow process at room temperature. Heating of a thin metal filament by an electric current produces the temperatures needed for a rapid reaction. For example, zirconium combines chemically with oxygen at 800°C to produce the white solid zirconium oxide, ZrO_2. Magnesium oxide, MgO, and aluminum oxide, Al_2O_3, are also white solids. The balanced chemical equations for these oxidation or combustion reactions are summarized as follows:

$$2 \text{ Mg (s)} + O_2 \text{ (g)} \rightarrow 2 \text{ MgO (s)}$$

$$\text{Zr (s)} + O_2 \text{ (g)} \rightarrow ZrO_2 \text{ (s)}$$

$$2 \text{ Al (s)} + 3 O_2 \text{ (g)} \rightarrow Al_2O_3 \text{ (s)}$$

KEY TERMS

combustion; oxidation

REFERENCES

1. Leggat, R. (1996). A history of photography: Lighting, http://www.kbnet.co.uk/rleggat/photo/history/lighting.htm
2. "Robert Wilhelm Bunsen (1811–1899)," http://www.woodrow.org/teachers/chemistry/modules/1992/Bunsen.html

RELATED WEB SITES

"Picture This . . . History of Photography," http://www.picturethis.net/history.html

WHY DID DOROTHY HAVE TO OIL THE TIN WOODMAN IN *THE WONDERFUL WIZARD OF OZ*?

"What can I do for you?" [Dorothy] inquired softly, for she was moved by the sad voice in which the man spoke.

"Get an oil-can and oil my joints," he answered. "They are rusted so badly that I cannot move them at all; if I am well oiled I shall soon be all right again."

—An excerpt from "Chapter 5: The Rescue of the Tin Woodman" in The Wonderful Wizard of Oz, *by L. Frank Baum*[1]

THE CHEMICAL ESSENCE

Had the Tin Woodman been constructed of pure tin he would not have rusted. Typically, the term *rust* is reserved for the product of the oxidation of the metal iron or its alloys, often due to atmospheric conditions. The Tin Woodman most likely was constructed of the same type of material used for "tin cans"—*tinplate*—a thin sheet of iron or steel (an iron alloy) coated with tin. The iron component of the Woodman's framework oxidized in air to produce the product iron oxide or rust.

THE CHEMICAL SPECIFICS

The oxidation of iron by oxygen is a spontaneous process in acidic, neutral, or basic environments. Iron serves as the *anode* and *reducing agent* as it is oxidized to Fe^{2+}:

$$Fe\ (s) \rightarrow Fe^{2+} + 2\ e^-$$

Oxygen acts as the *oxidizing agent*. The products of the reduction of O_2 depend on the acidity of the environment. In acidic solution, the reduction of O_2 generates water:

$$O_2 + 4\ H^+ + 4e^- \rightarrow 2\ H_2O$$

In alkaline or neutral solution, hydroxide ions are the oxidation product:

$$O_2 + 2\ H_2O + 4\ e^- \rightarrow 4\ OH^-$$

Thus, the overall reaction in acidic solution is as follows:

$$2\ Fe\ (s) + O_2\ (g) + 4\ H^+ \rightarrow 2\ Fe^{2+}\ (aq) + 2\ H_2O\ (l)$$

In a neutral or alkaline solution, the oxidation of iron is represented by the following:

$$2\ Fe\ (s) + O_2\ (g) + 2\ H_2O \rightarrow 2\ Fe^{2+}\ (aq) + 4\ OH^-\ (aq)$$

Thus, the primary reactions at the surface of iron are the loss of the metal due to oxidation to the divalent ion and the reduction of O_2 gas to either water or hydroxide ion. The formation of rust is actually a secondary oxidation reaction of the Fe^{2+} ions to Fe^{3+} with additional O_2, forming insoluble Fe_2O_3:

$$4\ Fe^{2+} + O_2\ (g) + 4\ H_2O \rightarrow 2\ Fe_2O_3 + 8\ H^+$$

Tin plating is a common procedure to protect iron and its alloys from rusting. Tin is less easily oxidized than iron, as revealed by the relative magnitudes of the standard reduction potentials of the two metals:

$$\text{Sn}^{2+} + 2 \text{ e-} \rightarrow \text{Sn (s) } \varepsilon° = -0.14 \text{ V}$$

$$\text{Fe}^{2+} + 2 \text{ e-} \rightarrow \text{Fe (s) } \varepsilon° = -0.44 \text{ V}$$

As long as the tin coating remains intact, the iron is protected from oxidation. A scratch of the tin surface to expose the iron metal leads to the oxidation of iron in preference to the tin.

KEY TERMS

oxidation; oxidizing agent; anode; reducing agent

REFERENCE

1. Baum, L. F. Chap. 5: The rescue of the Tin Woodman. *In: The Wonderful Wizard of Oz,* http://www.ukans.edu/carrie/kancoll/books/baum/oz05.htm

RELATED WEB SITES

"Tin and Tin Coatings," MFSA Quality Metal Finishing Guides, Technology in Review, Enthone-Omi Resource Center, Enthone-Omi, Inc., http://www.enthone-omi.com/TechRev/mfsa/tintin.html

WHY DO OLD PAINTINGS DISCOLOR?

Art conservation, particularly painting restoration, is an important endeavor to preserve our cultural heritage and maintain the aesthetic value of an artistic piece. Chemical reactions occurring on a microscopic level are the origin of the macroscopic changes that we observe as aging.

THE CHEMICAL ESSENCE

Artists incorporate a variety of pigments in their paints to provide the color necessary to express their creative intentions. To maintain the artist's colors, museums and art collectors know the importance of such environmental factors as temperature, relative humidity, light intensity, and air quality. Nevertheless, the nature of the work of art itself may influence the aging properties of the paint. The origin and purity of the pigments used, the combination of pigments selected, the pigment volume applied, and the type of binding medium utilized (e.g., drying oils such as linseed) are factors that influence the painting's permanence.[1] Throughout history, many painters have been well aware of the aging properties of their paints and took precautions to prevent these changes.[2] Nevertheless, unanticipated chemical reactions can occur between the materials in the painting and as a consequence of environmental exposure. In response to the new material environment that forms as the painting ages, additional changes in the physical and chemical properties of the pigments can subsequently occur to discolor the work further.

THE CHEMICAL SPECIFICS

A study of the chemical aspects of the aging of works of art is a vast undertaking. Many of the investigations focus on the artist's selection of pigments. In particular, a variety of lead-containing pigments have been used by artists throughout the centuries. Lead-containing white pigments include *carbonate white lead* used by the ancient Egyptians, Greeks, and Romans with a variable composition generally designated as $2\,PbCO_3 \cdot Pb(OH)_2$. *Basic lead sulfate* is a white pigment also of variable composition such as $PbSO_4 \cdot PbO$ or $(PbSO_4)_2 \cdot Pb(OH)_2$. *Chrome yellows* and *oranges* consist of various proportions of lead chromate, $PbCrO_4$ (the chief constituent), lead sulfate, $PbSO_4$, and lead monoxide, PbO. *Chrome greens* are mixtures of Prussian blue, $Fe_4[Fe(CN)_6]_3$ and lead chromate, $PbCrO_4$. *Molybdate oranges* with a red tendency are made from lead chromate precipitated together with lead molybdate, $PbMoO_4$ and lead sulfate, $PbSO_4$. *Red lead,* Pb_3O_4, is used extensively as a primer paint.

A commonly held belief is that lead-containing pigments react with hydrogen sulfide in polluted air to form the black precipitate lead sulfide. On exposure to hydrogen sulfide gas, all of these pigments will darken because of the formation of the black lead sulfide, PbS:

$$PbCO_3 \text{ (white)} + H_2S \rightarrow PbS \text{ (black)} + H_2O + CO_2$$

However, hydrogen sulfide in industrial atmospheres is rapidly oxidized to sulfur dioxide and sulfuric acid.

Alternative explanations for the discoloration of old paintings are available. For example, exposure of lead-based pigments to sulfide-based pigments will also contribute to the darkening over time. Specifically, white lead and chrome yellow (lead-containing pigments) should not be used with the blue pigment *ultramarine,* a complex sodium aluminum silicate and sulfide found in the mineral *lapis lazuli.* The key ingredients for the formation of black lead sulfides are present when combinations of these dyes are used.

Other chemical factors can cause pigments to darken. Cuprous oxide, Cu_2O, starts out as bright red pigment, but it gradually oxidizes to the cupric form, CuO, which is characteristically black. Prolonged exposure to light can cause copper resinate, a transparent green pigment used in the 15th to 18th centuries, to become a deep chocolate brown. Prussian blue, $Fe_4[Fe(CN)_6]_3$ or ferric ferrocyanide, should not be used with basic carbonate white lead or other basic pigments because of the reaction to precipitate ferric hydroxide, imparting a reddish tinge to the paint:[3]

$$Fe^{3+}(aq) + 3\ OH^-\ (aq) \rightarrow Fe(OH)_3\ (s)$$

KEY TERMS

oxidation; precipitation

REFERENCES

1. "Dosimetry of the Museum Environment: Environmental Effects on the Chemistry of Paintings," Oscar F. van den Brink, Molecular Aspects of Ageing in Painted Works of Art, Progress Report 1995–1997, FOM Institute for Atomic and Molecular Physics, Amsterdam, NL, http://www.amolf.nl/departments/molart/progress97/progress97.html#Dosimetry_of_the_Museum_Environment
2. Molecular Aspects of Ageing in Painted Works of Art, Progress Report 1995–1997, FOM Institute for Atomic and Molecular Physics, Amsterdam, NL, http://www.amolf.nl/departments/molart/progress97/progress97.html
3. Stevens, K. K., and Warner, J. C. (1953). Organic protective coatings: Paints, varnishes, enamels, lacquers. *In: Chemistry of engineering materials,* 4th ed. (J. C. Warner, ed.), pp. 544–590. McGraw-Hill, New York.

RELATED WEB SITES

"Cadmium Pigments," J.R.J van Asperen de Boer, Molecular Aspects of Ageing in Painted Works of Art, Progress Report 1995–1997, FOM Institute for Atomic and Molecular Physics, Amsterdam, NL, http://www.amolf.nl/departments/molart/progress97/progress97.html#Cadmium_pigments

"A Fine Art: Conservation Chemists Go Ga-Ga Over Great Works," Mark Uehling, http://www.acs.org/memgen/rxntimes/rxt0297/artfldr.html

"Orpiment, Deterioration of Arsenic Sulfide Pigments," Arie Wallert, Molecular Aspects of Ageing in Painted Works of Art, Progress Report 1995–1997, FOM Institute for Atomic and Molecular Physics, Amsterdam, NL, http://www.amolf.nl/departments/molart/progress97/progress97.html#Orpiment_deterioration_of_arsenic_s

"Science and Art Broadcast," The Chicago Academy of Sciences, http://www.caosclub.org/
    archive/art1.html
"A Team from the Universitat Politecnica de Catalunya Verify the Authenticity of a Sketch
    by Francisco de Goya Recently Found in Palafrugell (Girona)," Universitat Politecnica
    de Catalunya, Archive, Past Research and Institutional News: Materials Technology News,
    http://www.upc.es/english/noticies/noticies.htm
"Yellow Lake Pigments, Identification and Study of Deterioration of Flavonoid Colorants,"
    Arie Wallert and N. Wyplosz, Molecular Aspects of Ageing in Painted Works of Art,
    Progress Report 1995–1997, FOM Institute for Atomic and Molecular Physics, Amsterdam,
    NL,   http://www.amolf.nl/departments/molart/progress97/progress97.html#Yellow_lake_
    pigments_identification

REFERENCE

Zieske, F. (1995). An investigation of Paul Cezanne's watercolors with emphasis on emerald
    green. *In: The book and paper group annual,* Vol. 14. Philadelphia Museum of Art, http://
    sul-server-2.stanford.edu/aic/bpg/annual/v14/zieske.html

WHAT CAUSED THE DESTRUCTION OF THE ORGAN PIPES IN THE CATHEDRALS OF NORTHERN EUROPE DURING THE 19TH CENTURY?

*Restorations of 19th century organs in the cathedrals of northern Europe
revealed a metal "disease" often attributed to the corrosion of tin. Chemically
speaking, however, the structural change in the metal pipes is a completely
different phenomenon. What aspects of chemistry must organ builders consider when attempting to achieve a particular acoustical character?*

THE CHEMICAL ESSENCE

Organ pipes today are made of a variety of woods (e.g., mahogany)
and metal alloys depending on the desired tonal quality, appearance,
and cost. Most pipes are made of a varying mixture of tin and lead
called spotted metal, but pipes with copper and zinc are also common.[1]
It is widely believed that the higher the tin content, the brighter the
tone and the shinier the appearance. Historically, tin was the preferred
material for an organ pipe, yet the expense of the metal was always a
consideration and a limitation.

Pure tin exhibits two common forms in the solid state—a gray tin
and a white tin. At temperatures above 13°C or 55°F, the more stable
form of tin is the denser white tin. At lower temperatures, the white

tin is slowly converted to the gray form, a more powdery substance. Prolonged exposure to the cold winter temperatures of northern Europe contributed to the loss of integrity and disintegration of many cathedral organ pipes. As a consequence of the progressive nature of the structural transformation, as the white tin metallic surface becomes covered with gray powder, the degradation is often known as "tin disease," "tin pest," or "tin plague."

THE CHEMICAL SPECIFICS

At atmospheric pressure, pure solid tin adopts two structures or *allotropes* depending on temperature. At room temperature white metallic tin is stable but, at temperatures below 13°C, white tin undergoes a phase transformation into gray tin. White tin (also known as β-tin) adopts a body-centered tetragonal crystal structure (Fig. 2). Allotropic gray tin (α-tin) crystallizes in a cubic diamond crystal structure (Fig. 3).

KEY TERMS

allotrope; phase change; phase diagram

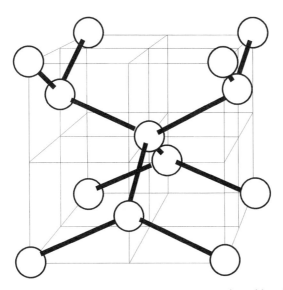

Figure 2   A body-centered tetragonal crystal structure adopted by white tin.

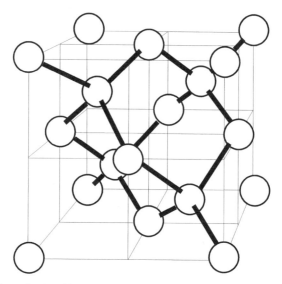

Figure 3   A cubic diamond crystal structure adopted by gray tin.

REFERENCE

1. "Front Pipes, Principals, Mixtures & Mutations, Flutes and Strings," F. J. Rodgers Ltd., Craftsman Organ Pipe Makers and Voicers, http://www.musiclink.co.uk/pipeorgan/flues. html

RELATED WEB SITES

"Behind the Pipes," D. A. Flentrop, http://www.chapel.duke.edu/organs/Flentrop/ memorial/flentrop.htm
"Materials," David Harrison, School of Science & Technology, Athrofa Addysg Uwch Gogledd Ddwyrain Cymru, North East Wales Institute of Higher Education, Wrexham, North Wales, http://www.newi.ac.uk/buckleyc/materials.htm
"The Science Corner: Allotropes," Nigel Bunce and Jim Hunt, College of Physical Science, University of Guelph, http://www.physics.uoguelph.ca/summer/scor/articles/scor40.htm

HOW DO FOG MACHINES CREATE THE ARTIFICIAL FOG OR SMOKE USED IN THEATRICAL PRODUCTIONS?

Whether for a dramatic scene or a science fiction fantasy, the use of artificial fog to create ambiance and special effects is commonplace in

Figure 4   The chemical structure of propylene glycol, used in artificial fog machines.

theatrical and film productions. On what basic chemical principles do fog machines operate?

THE CHEMICAL ESSENCE

Some of the most memorable scenes in a motion picture are the creation of special effects technicians. One standard tool in the entertainment industry to create ambiance, simulate specific designs such as volcanoes or swamps, or accentuate optical effects is fog production. (In the theatrical sense, the terms *fog* and *smoke* are used interchangeably for mist consisting entirely of liquid droplets. In the chemical sense, *fog* refers to a liquid phase dispersed in a gas, whereas *smoke* contains solid particulate matter dispersed in a gas.) A variety of approaches are used to create artificial fog or smoke.

Many companies sell machines that use a glycol- and water-based fog fluid.[1,2] These liquids are pumped into a temperature-controlled heat exchanger, heated, and, as a consequence, vaporized into thick clouds of fog. Different percentages of propylene glycol (Fig. 4) and triethylene glycol (Fig. 5) help to achieve the desired effect, for example, by creating an invisible particulate mist that enhances the viewing of laser lightbeams or by concocting a dense, white fog with varying "hang times." These thermally evaporated glycol-based conventional fog machines are preferred over "cracked-oil" machines that rely on atomization produced by pressurization.[3] Both enhanced fire safety and reduced residue deposits despite faster dissipation times are characteristic of the thermal evaporation machines.

A number of health concerns have been raised with glycol-based fog machines.[4] In particular, the *hygroscopic* nature (i.e., tendency to absorb

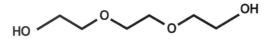

Figure 5   The chemical structure of triethylene glycol, used in artificial fog machines.

moisture) of glycols is purported to cause a drying effect on the nose, eyes, and throat. Some manufacturers have designed methods for creating superfine fog droplets solely from water without excessive wetting. The Academy of Motion Picture Arts and Sciences Board of Governors granted a Technical Achievement Award in 1998 to three employees of Praxair, Inc., for their creation of an artificial fog that uses a mixture of Praxair oxygen, nitrogen, and steam.[5,6]

How can one keep the fog low to the ground? Chilling the fog will help. Lowering the temperature of the fog causes the density of the fog to increase and the seemingly heavier mist will sink to lower levels. Some foggers utilize dry ice or liquid CO_2 to produce a low-lying fog effect.[7] These refrigerants are used to cool a manifold through which the fog passes.

As an interesting application of fog-making technology, companies have designed *fog security systems.* These security systems generate a dense fog to reduce visibility to zero and trap would-be thieves in the fog until authorities arrive on the scene.[8,9]

THE CHEMICAL SPECIFICS

Why are mixtures of polyfunctional alcohols (i.e., containing more than one hydroxyl $-OH$ group) used to produce smoke and fog effects? As we have seen, small liquid particles dispersed in a gas create a fog. Initially, the mixture of propylene glycol, triethylene glycol, and water is liquid. The resulting fog created by vaporizing a mixture of these liquids is dense, white, and odorless. In actuality, primarily the water in these mixtures is vaporized (boiling point of 100°C), whereas propylene glycol (boiling point of 187.6°C) and triethylene glycol (boiling point of 285°C) remain largely in the liquid phase. The *hygroscopic* nature of propylene glycol and triethylene glycol arises from their ability to form strong hydrogen bonds and ensures the favorable dispersion of small droplets of these liquids in the water vapor. The white color of fog, just like the white color of clouds, is a consequence of the equal scattering of all wavelengths of light by the suspended liquid droplets.

KEY TERMS

fog; smoke; hygroscopic; hydrogen bonding; polyfunctional alcohol

REFERENCES

1. B. T. Productions, http://www.btprod.com/fogfaq.html
2. "Atmospheres Lighting Enhancement Fluid," HighEnd Systems, Lightwave Research, http://info.highend.com/hes/FOG/juice/atmos.html

3. "A Comparison between Thermally Evaporated Glycol-Based Fog Machines and Atomization by Pressurization of Cracked Oil," HighEnd Systems, Lightwave Research, http://info.highend.com/hes/FOG/belfog/belfog.html

4. "Glycol, Glycerin, and Oil Fog Reading List," Entertainment Services & Technology Association, http://www.esta.org/fogdocs.htm

5. "New, Safer Fog Wins Academy Award," *Inside Science,* American Institute of Physics, http://www.aip.org/inside_science/scripts/90.htm

6. "Academy Awards Praxair Employees Technical Achievement Certificate," Praxair, http://www.praxair.com/Praxair.nsf/2b8c767ac26ecc5b8525654400157fde/cabd4f9cc57eccbe852565b00075aa23?OpenDocument

7. "Coldflow LCO2 Exchanger Module," HighEnd Systems, Lightwave Research, http://info.highend.com/hes/FOG/cflow1/cfcopy.html

8. "Smokecloak," National Tint & Security Ltd., http://www.smokecloak.ca/sm2.html

9. "Development Concept behind Fog Security Systems," Fog Security Systems, Inc., http://www.tbf.net/fog/concpt.htm

RELATED WEB SITES

"Smoke without Fire," A Historical Look at Fog, by Mike Wood, HighEnd Systems, Lightwave Research, http://info.highend.com/hes/FOG/swf/swf.html

"Special Effects Fog," Mee Industries, Inc., http://www.meefog.thomasregister.com/olc/meefog/mee01.htm

Chapter 9

Connections to Currency and Gems

WHY IS IT INCORRECT TO CALL U.S. CURRENCY "PAPER CURRENCY"?

We all recognize the Americans whose portraits appear on the front face of U.S. currency: George Washington ($1), Thomas Jefferson ($2), Abraham Lincoln ($5), Alexander Hamilton ($10), Andrew Jackson ($20), Ulysses Grant ($50), Benjamin Franklin ($100). As with U.S. coinage, the Secretary of the Treasury in consultation with the Commission on Fine Arts selects the designs shown on U.S. currency. However, it is the Bureau of Engraving and Printing that is responsible for designing and printing U.S. currency at its facilities in Washington, D.C., and Ft. Worth, Texas. Although the printed features of U.S. currency have undergone significant changes over the years, the chemistry of the "paper" has a much more stable history.

THE CHEMICAL ESSENCE

The constant circulation of a national currency demands a durable, yet high-quality material. Beginning with the first series of U.S. banknotes issued in 1861, U.S. currency has never been printed on paper but on a cotton/linen fabric with the linen content held at 25 ± 5%. Often this fabric is referred to as cotton and linen rag paper. Red and blue silk fibers from scraps and cuttings of clothing manufacturers are embedded in the cotton/linen sheet to deter counterfeiters. A commercial company, Crane & Company, Inc., of Dalton, Massachusetts, has been manufacturing the rag paper since 1879, and it is illegal for anyone to manufacture, possess, or use this material or a fabric of a similar type.

The history of Crane & Company is a fascinating story.[1] The association of the Crane family with U.S. currency began with the sale of paper in 1775 by Steven Crane to Paul Revere for the Massachusetts Bay Colony's first currency. Having been taught the art of paper making by his father, Zenas Crane carried on the family business and built his first paper mill along the Housatonic River in Dalton, Massachusetts, in 1801. The following advertisement appeared in the *Pittsfield Sun* in 1800:[2]

> Americans:
> Encourage your own manufactories, and they will Improve. Ladies save your Rags. As the Subscribers have it in contemplation to erect a Paper-Mill in Dalton, the ensuing Spring; and the business being very beneficent to the community at large, they flatter themselves they shall meet with due encouragement.
> [Signed]
>
> —Henry Marshall, Zenas Crane, and John Willard

As this notice indicates, cloth rags were the basic raw material for conversion into pulp for high-quality rag-based paper products. Although Zenas Crane retired in 1842, his sons James Brewer Crane and Zenas Marshall Crane continued the business and began to make paper for banknotes, bonds, and securities. Two generations later in 1879, W. Murray Crane won a competition to manufacture the paper for U.S. currency. Crane & Company has held an exclusive contract with the United States Treasury Department to produce the specially threaded cotton and linen paper for U.S. currency. The company also manufactures paper for the foreign currencies of Canada, Mexico, Indonesia, and the Ukraine,[3] in addition to its renown high-quality stationery business.

THE CHEMICAL SPECIFICS

In ancient Egypt the fibers and glue-like sap of the reedy papyrus plant were the main constituents of the sheets formed for writing materials. Other woody fibers, such as mulberry, were introduced by the Chinese in the first century A.D. Paper mills existed in Europe by the 14th century, and linen and cotton rags served as the basic raw materials through the 18th century. Paper mills often solicited publicly for rags because the shortage of raw materials could not keep up with the demand for paper. The manufacturing of paper from wood pulp began in 1800 to relieve the paper industry from its demand for cotton and linen rags. Even today grades of paper requiring strength, durability, permanence, and fine texture employ cotton and linen fibers derived from textile and garment mill cuttings.

One of the final steps in the paper making process is the coating of the paper surface to achieve a variety of effects. For example, coating can enhance the uniformity of the surface for printing inks or enhance the opacity or whiteness of the paper. Titanium dioxide (TiO_2) is used to whiten and opacify all U.S. paper currency and many other forms of paper.[4] Why is titanium dioxide an optimal coating selected to whiten paper? For the human eye to perceive the color white, an object must scatter all wavelengths of light. Thus, the ability of a sheet of paper, or its coating, to scatter light will define its opacity and brightness. One mechanism of light scattering is by *refraction,* and the high refractive index of titanium dioxide makes it an effective scatterer of light. Refraction is defined as the bending of light as it passes from one medium, such as air, to another medium, such as a paper coating. The larger the difference in the refractive index of the two media, the greater the extent to which the light is bent or refracted. The extremely high refractive index values of both commonly used crystal forms of titanium dioxide—anatase and rutile—make them ideal commercial pigments for achieving brightness with low volumes of sample.[5]

KEY TERMS

Refraction

REFERENCES

1. Furash, M. The state of one small family business: Crane & Co. *Inc. Online,* http://www.inc.com/incmagazine/archives/27961141.html
2. "Crane Museum, Dalton, Massachusetts 01226," http://www.berkshireweb.com/themap/dalton/dalton.html
3. Donn, J. Opening up money supply worries Dalton paper firm. Associated Press, *The Standard-Times,* http://www.s-t.com/daily/04-97/04-13-97/f03bu299.htm
4. "Millenium Inorganic Chemicals Fun Facts," http://www.mic-global.com/mi_htmlcode/mi_use/mi_use.html#top
5. "DuPont Ti-Pure: Paper: Rutile and Anatase," http://www.dupont.com/tipure/paper/anarut.html

RELATED WEB SITES

"Anatomy of a Bill: The Currency Paper, Secrets of Making Money," NOVA Online, http://www.pbs.org/wgbh/nova/moolah/anatomypaper.html

"A Brief History of Our Nation's Paper Money," 1995 Annual Report, Federal Reserve Bank of San Francisco, http://www.frbsf.org/frbsf/pubs/annualrpt/history.html

"Crane Bro's Paper Manufacturers," http://hamp.hampshire.edu/~laoGU/hampden/p157.html

"Currency: A History of U.S. Money," Federal Reserve Bank of Kansas City. http://www2.minnbankcenter.org/mba/frnotes.htm

"DuPont Ti-Pure titanium dioxide," http://www.dupont.com/tipure/paper/index.html

"Fundamental Facts about U.S. Money," Federal Reserve Bank of Atlanta, http://www.atl.frb.org/publica/brochure/fundfac/money.htm

"History of Paper Money," Federal Reserve Bank of San Francisco, http://www.frbsf.org/econedu/currencyex/funfacts.html#A3

"The Making of Money," Bureau of Engraving and Printing, United States Treasury, A 39-second video with sound on the currency-making process. http://www.treas.gov/bep/making.mpg

"The Mineral Anatase,"Amethyst Galleries, Inc., http://mineral.galleries.com/minerals/oxides/anatase/anatase.htm

"The Mineral Rutile,"Amethyst Galleries, Inc., http://mineral.galleries.com/minerals/oxides/rutile/rutile.htm

"Money Hot Off the Press," Bureau of Engraving and Printing, United States Treasury, A 30-second video capsule of how currency is printed. http://www.treas.gov/bep/money.mpg

"Money Making," The AFU and Urban Legend Archive, http://www.urbanlegends.com/misc/money_making.html

"Six Kinds of United States Paper Currency," Kelley L. Ross, Ph.D., http://www.friesian.com/notes.htm

"Treasury Home Page: Existing Contracts: Massachusetts," http://www.treas.gov/sba/ma.html

OTHER INTERESTING REFERENCES

Hallett, A., and Hallett, D. Zenas Crane. (1997). _In: Encyclopedia of Entrepreneurs,_ pp. 131–132. John Wiley, New York.

WHAT IS THE PURPOSE OF THE THREAD THAT RUNS VERTICALLY THROUGH THE CLEAR FIELD ON THE FACE SIDE OF U.S. CURRENCY?

The Bureau of Engraving and Printing in Washington, D.C. is responsible for the design, engraving, and production of U.S. banknotes. Several of the special design features that are evident in U.S. currency today take advantage of technological advancements in the ink and paper industries. The highly sophisticated chemistry of such materials is an effective deterrent against counterfeiting.

THE CHEMICAL ESSENCE

How closely have you examined the details on paper currency? The denomination, portrait, and back design are commonly recognized. The

Treasury seal, Federal Reserve seal, signature of the U.S. treasurer, motto, and serial numbers are also clearly identified. Some design features are not as readily apparent. One such feature introduced in 1990 series currency is a clear thread embedded in paper currency as a security measure against counterfeiting. This security thread is made of polyester and has a denomination identifier printed on it. Both the thread and the printing are visible only with a light source and can be viewed from either the face or the back of the note. For the two highest denomination bills, $100 and $50, the security thread repeats the wording "USA 100 USA 100" and "USA 50 USA 50," respectively. For the next three lower denominations ($20, $10, and $5), the printing consists of the abbreviation "USA" followed by the written denomination in capital letters, as in "USA TEN," repeated along the length of the thread. No security thread is included in the $1 note. Beginning with the 1996 series notes, the placement of the thread varies, with the position indicative of the bill's denomination. In addition, the thread on the new $100 note first issued on March 25, 1996, was redesigned to glow or fluoresce a red color when exposed to ultraviolet light in a dark setting. The thread is positioned to the immediate left of the oval frame on Ben Franklin's portrait. The new $50 note first issued on October 27, 1997, also has the added feature that the security thread, now to the right of the portrait of President Grant, glows yellow when exposed to ultraviolet light in a dark environment. The printing on the thread also includes a flag in addition to "USA 50." The redesigned $20 bill was unveiled in May 1998 and was put into circulation in the fall of that year. The vertical thread is embedded to the far left of President Jackson's portrait (to the left of the Federal Reserve Seal). The words "USA TWENTY" and a flag can be seen from both sides against a light. The number "20" appears in the star field of the flag. The thread fluoresces green under an ultraviolet light. Design features for the new $10 note, to be released in the summer of 1999, have not yet been revealed.

The introduction of the new security measures has not always occurred without error. In November 1996 officials at the Bureau of Printing and Engraving of the Treasury Department discovered that the positions of the polymer security thread and the watermark on $4.6 million worth of the newly designed $100 bills were swapped—the thread incorrectly appeared on the right side of Benjamin Franklin's portrait and the watermark on the left.[1] These misprinted bills are still legal tender, yet they are likely to be worth much more than their face value to collectors. You might wonder why the U.S. Treasury is varying the position of the security thread in its currency. Counterfeiters often attempt to "raise notes," that is, bleach out the paper of a low denomination and reprint a higher denomination onto the authentic paper. With the position of the security thread constrained to the bill's denomination, this type of fraud will be easily detected.

The Bank of Latvia (Latvijas Banka) also incorporates invisible fluorescent fibers in its currency as a security measure. As with the security thread on United States banknotes, the Latvian thread fluoresces under ultraviolet light, emitting three different colors. In addition, the printing ink employed for the red serial numbers located in the upper center portion of the face of the notes fluoresces upon exposure to ultraviolet light. Singapore currencies include specially formulated inks to fluoresce under ultraviolet light, including black serial numbers that glow green and a red seal of the minister of finance that emits orange illumination. Fluorescent inks are employed for denomination figures, seals, and other objects in Hong Kong dollars, Malaysian dollars, Taiwan yuan, and Indonesian rupiahs. Special fluorescent fibers are incorporated in the paper-making process for Deutsch marks, Italian lire, Netherlands goldens, and Belgian francs. Fluorescent spots also appear on Swiss francs and Canadian dollars. A number of commercial products are manufactured to detect embedded fluorescent denomination-specific security threads.[2]

THE CHEMICAL SPECIFICS

The security thread introduced in the series 1990 banknotes is a thin metallized polyester strip that is 1.4 to 1.8 mm in width and 10 to 15 μm in thickness.[3] Polyester threads are standard synthetic fibers consisting of large linear (chainlike) or cross-linked (network) polymers formed from a large number of smaller molecules or monomers. The monomers are connected via *ester linkages* as in Figure 1. Polyesters most commonly are prepared from equivalent amounts of two different monomers: *glycols* and *dibasic acids*. Glycols are organic compounds containing two hydroxyl groups, $-OH$, and dibasic acids are organic molecules containing two carboxyl functionalities, $-COOH$.

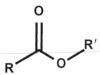

Figure 1   Ester linkages that connect monomers to form polymeric molecules known as polyesters.

KEY TERMS

fluorescence; polyester; ester linkage; glycol; dibasic acid

REFERENCES

1. Officials report $100 printing error. (1997, March 29). *National/World News,* TH On-Line, Telegraph Herald of Dubuque, Iowa, http://www.thonline.com/th/news/032997/National/ 52409.htm
2. "Ultraviolet Applications," UVP Products, Inc., Upland, CA, http://www.uvp.com/html/ bulletins.html
3. Description and assessment of deterrent features. (1993). *Counterfeit deterrent features for the next-generation currency design,* chap. 4. National Materials Advisory Board, National Academy Press, Washington, D.C.

RELATED WEB SITES

"Counterfeit Money Detector," Trade2000 (USA), Inc., http://www.trade2000.com/intro.htm and http://www.trade2000.com/detector3.htm
"Features of the New Twenty," Bureau of Engraving and Printing, http://www.bep.treas.gov/ currency/20features.cfm
"Know Your Money," United States Secret Service, United States Treasury, http://www.treas. gov/kids/money/kymintro.html
"The Making of Money," Bureau of Engraving and Printing, United States Treasury, A 39-second video with sound on the currency-making process, http://www.treas.gov/bep/ making.mpg
"Money Hot Off the Press," Bureau of Engraving and Printing, United States Treasury, A 30-second video capsule of how currency is printed, http://www.treas.gov/bep/money.mpg
Phosphors for Security, Tagging and UV/IR Detection, Phosphor Technology, http://www. phosphor.demon.co.uk/iruv.htm
"Physical Sciences and Technology. New greenbacks: How to make a buck—literally," Richard Lipkin, http://www.sciencenews.org/sn_edpik/ps_4.htm
"Security Features of the Lats Banknotes," Latvijas Banka, http://www.bank.lv/naudas/ English/index_security.html
U.S. Treasury and Federal Reserve introduce new $50 bill. (1997, June 12). *Treasury News,* http://www.treas.gov/press/releases/pr1746.htm
"Your Money Matters: Germany. Description of the DEM 100," http://users.hol.gr/~friar/ dem100.html

WHY DOES THE DENOMINATION ON THE LOWER RIGHT-HAND CORNER OF NEW U.S. CURRENCY SHIFT IN COLOR FROM GREEN TO BLACK WHEN VIEWED FROM DIFFERENT ANGLES?

After nearly four generations, the currency of the United States has undergone a noticeable change in appearance using several technological advances.

In particular, complex and innovative chemistry has been incorporated in the design of the ink used to denote the denomination in the lower right-hand corner. Is it an optical illusion, or is the color of the ink shifting from green to black as you tilt the bill?

THE CHEMICAL ESSENCE

With advancements in the technologies of color copiers, scanners, and printers, maintaining the security of U.S. currency is increasingly difficult. Since 1990 the U.S. Treasury has added a number of features to new currency as security measures against counterfeiting. One of the new design features for series 1996 notes is the use of *optically variable ink* on the number in the lower right-hand corner of the bill. As one tilts the bill in light, the use of a color-shifting ink becomes apparent. When viewed straight on, the numerals appear green, but when viewed at an angle, the numbers look black.

The technology was developed by chemist Roger Phillips, currently product technology manager and manager of intellectual properties at Flex Products in Santa Rosa, California. Phillips and two colleagues, Pat Higgins and Peter Berning at Optical Coating, originally designed an optically variable foil that would be applied to U.S. currency as an anticounterfeiting measure. The Federal Reserve initially liked the idea and spent $17 million to develop the product, but by 1985 federal officials decided to abandon the new technology because the process would require expenditures for new machinery. With the company facing massive layoffs, Phillips came up with the idea to translate the technology into a printing, which would not require costly new equipment.

The United States is not the only government using optically variable ink as a security feature. More than 40 countries use the pigment technology. On the face side of the banknotes issued by the Banque de France, optically variable ink shifting in color from green to blue is used for the figures representing the subatomic world on the Pierre and Marie Curie 500 franc note, for the cross-section of one of the four pillars of the Eiffel Tower on the Gustave Eiffel 200 franc note, for Cezanne's palette on the Paul Cezanne bill for 100 francs, and for the "boa digesting an elephant" on the Saint-Exupery 50 franc bill. On the 500-lats note issued by the Bank of Latvia (Latvijas Banka), optically variable ink creates the effect of changing colors for the nominal value printed in the left-hand corner of the obverse side of the bill. Security measures involving optically variable ink have also been taken by the Central Bank of Ireland for its banknotes. The United Kingdom also recently used optically variable ink on first-class rate stamps reissued in gold to commemorate the 50th anniversary of the marriage of Her Majesty the Queen to Prince Philip. The queen's head changes from gold to green with a change in viewing angle.

Roger Phillips has also extended the technology of optically variable ink to new areas, for example, revolutionizing the automotive industry. In 1993 he developed a durable automobile paint that changed colors when viewed from different angles.

THE CHEMICAL SPECIFICS

The process for developing optically variable ink involves the layering of several extremely thin metal-containing pigment coatings of precise thickness, followed by grinding of the coating into tiny platelets or flakes. The flakes are typically 1 μm thick and 2 to 20 μm in diameter with an average aspect ratio (i.e., ratio of width to height) of 10 to 1.[1] These color-shifting thin-film flakes are then suspended in a mixture of regular ink, and the ink is then applied to a surface. The high aspect ratio helps align the flakes parallel to the surface of the ink. As light shines on the flakes, light is reflected. Because of the random positioning of the metallic platelets, certain wavelengths of light are selectively reinforced ("constructive inter-ference"), whereas other wavelengths are canceled ("destructive interfer-ence"). This phenomenon, known as color diffraction, creates the appear-ance of color through reflection (see Fig. 2).

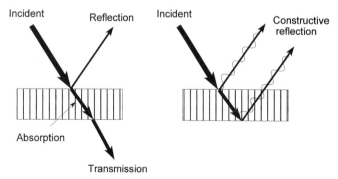

Figure 2 The phenomena of light reflection, absorption, and interference to create the appearance of color. Color arises as certain wavelengths of reflected light are selectively reinforced through constructive interference, while other wavelengths of light are canceled through destructive interference.

Partially reflective	Al	Cr	MoS_2	Fe_2O_3	Fe_2O_3
Low refraction	SiO_2	MgF_2	SiO_2	SiO_2	SiO_2
Inner reflector	Al	Al	Al	Al	Fe_2O_3
Low refraction	SiO_2	MgF_2	SiO_2	SiO_2	SiO_2
Partially reflective	Al	Al	MoS_2	Fe_2O_3	Fe_2O_3

Figure 3 Layering patterns of absorber (partially reflected)/dielectric (low refractive)/ reflector (inner reflector)/dielectric/absorber used to create the optical effect of color-shifting ink.

The particular colors that are observed at different angles depends critically on the thickness of the thin film coating. Precision instrumentation is required to carefully control film thickness during production. The magnitude of the optical effect depends on the density of flakes in the ink, while the quality of the optical effect depends on the precise orientation or alignment of these flakes with respect to the paper surface.

What are the typical materials used to create the color-shifting flakes? A symmetrical layering pattern of absorber/dielectric/reflector/dielectric/ absorber is employed, because the flakes can be oriented either up or down on the ink surface. The role of the *absorber* is to absorb particular wavelengths of light to enhance or modify the observable color change of the ink as the viewing angle is varied. The *dielectric* is a material that does not absorb visible light and also posses a low refractive index. The interference colors of such materials are highly angle dependent.[2] The *reflector* is the material that produces the unique optical effect through the reflection of light. Typical materials used in the multilayer structure are chromium as the absorber, magnesium fluoride or silicon dioxide as the dielectric, and aluminum as the reflector: $Cr/MgF_2/Al/ MgF_2/Cr$ and $Cr/SiO_2/Al/SiO_2/Cr$. Layer thicknesses of 50 Å, 4000 Å, 900 Å, 4000 Å, and 50 Å, respectively[1], are typical. Some additional combinations of materials for optically variable flakes are illustrated in Figure 3.

KEY TERMS

reflection; absorber; dielectric; reflector; ink

REFERENCES

1. Description and assessment of deterrent features. (1993). *In: Counterfeit deterrent features for the next-generation currency design,* chap. 4. National Materials Advisory Board, National Academy Press, Washington, D.C.
2. "Luster Pigments with Optically Variable Properties," Raimund Schmid, Norbert Mronga, Volker Radtke, Oliver Seeger, BASF Corporation, http://www.coatings.de/articles/schmid/schmid.htm

RELATED WEB SITES

"Bank Notes and Security Features," Banque de France, http://www.banque-france.fr/us/billets/secu/index.htm
"Know Your Money," United States Secret Service, United States Treasury, http://www.treas.gov/kids/money/kymintro.html
"The Making of Money," Bureau of Engraving and Printing, United States Treasury, A 39-second video with sound on the currency-making process. http://www.treas.gov/bep/making.mpg
"Money Hot Off the Press," Bureau of Engraving and Printing, United States Treasury, A 30-second video capsule of how currency is printed. http://www.treas.gov/bep/money.mpg
"Secrets of Making Money," NOVA Online, http://www.pbs.org/wgbh/nova/moolah
"Security Features of the Lats Banknotes," Latvijas Banka, http://www.bank.lv/naudas/English/index_security.html
"Stories from a Few Who Teach, Innovate, Create" by Mary Fricker, The Press Democrat, Outlook Sonoma County and Its Economy, http://www.pressdemo.com/outlook97/tech/stories.html
"United States Treasury $100 Information Page," "Color-Shifting Ink," http://www.treas.gov/currency/shift.html

HOW DO FORENSIC CHEMISTS USE VISIBLE STAINS TO TRAP THIEVES?

Thief detection powder is designed for thief detection and the identification of stolen or altered items. Once a marked article is touched, the powder creates a highly colored visible stain on the skin. What chemistry do forensic specialists apply to investigate crimes using detection powders?

THE CHEMICAL ESSENCE

Increasingly law enforcement agencies, banks, businesses, corporate offices, and concerned citizens are using detection powders to trap thieves. Thief detection powders are designed to be applied to the surface of an object likely to be stolen such as cash boxes, cash, or sensitive documents. Alternatively, the powder may be applied to doorknobs or tools to detect entry or use. When the powder comes in contact with the skin, a reagent in the powder reacts with the body's amino acids to create a highly visible long-lasting purple-colored stain. Although the stain will not immediately wash off the skin, the color will gradually rub off after about a day.

THE CHEMICAL SPECIFICS

Detection powders and fingerprint development kits commonly contain cream- or yellow-colored ninhydrin crystals or a solution of dissolved ninhydrin. Ninhydrin (also known as 1,2,3-indantrione, monohydrate; 2,2-dihydroxy-1,3-indandione; triketohydrindene, monohydrate; and triketohydrinden hydrate) has the structure presented in Figure 4. Ninhydrin will react with a free alpha-amino group, $-NH_2$. This group is contained in all amino acids, and analysis with ninhydrin is often performed to verify the presence of amino acids. When α-amino acids (i.e., amino acids with the structure $NH_2-CHR-COOH$) react with ninhydrin, a characteristic deep blue or purple color of reduced ninhydrin is observed. The reactions involved in this oxidation-reduction process are shown in Figure 5.[1]

KEY TERMS

α-amino acid; amine group

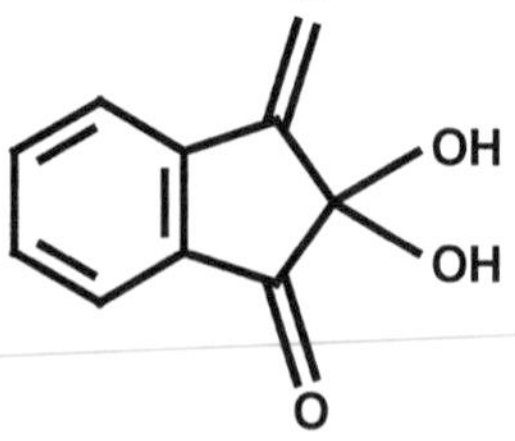

Figure 4   The chemical structure of ninhydrin.

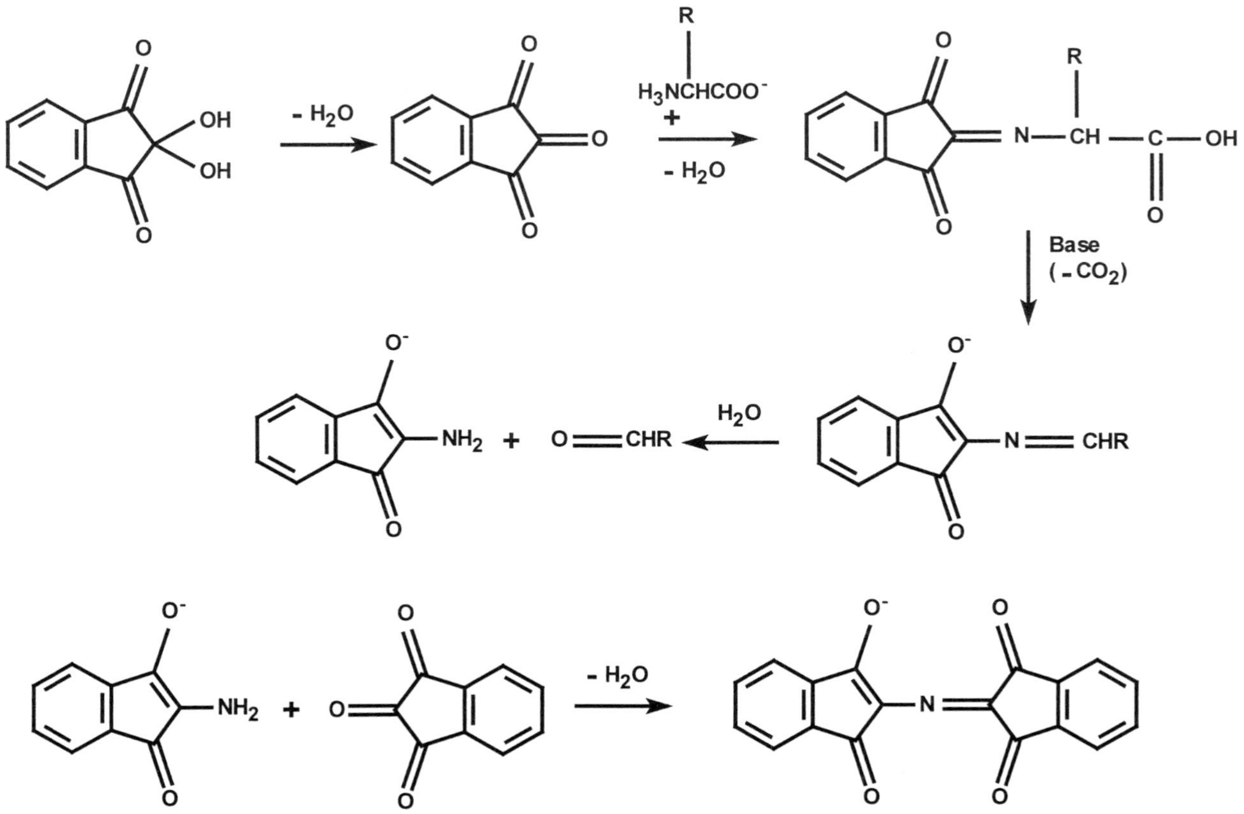

Figure 5 The sequence of reactions that reduce ninhydrin to produce a blue color.

REFERENCE

1. "Experiment 11. Proteins and Carbohydrates. Isolation of Casein and Lactose from Milk," McMaster University, Chem2006 Lab Manual, http://www.chemistry.mcmaster.ca/~chem2o6/labmanual/expt11/2o6exp11.html

RELATED WEB SITES

"The Art and Science of Criminal Investigation: Ninhydrin Processing," *Crimes and Clues,* Pat A. Wertheim, http://crimeandclues.com/ninhydrin.htm
"Visible Stain Thief Detection Powder," The Spy Store, http://www.spy-store.com/Theftpowders1.html
"Visible Thief Detection Compounds," Spy Headquarters, Sophisticated Security Solutions, http:/www.spyhq.com/pol3.htm

WHY IS THE HOPE DIAMOND BLUE?

Diamonds are the only gemstones whose colorlessness enhances their value. However, the rare, rich, natural coloring of "fancy color" diamonds commands the highest prices. The Hope Diamond possesses exceptional blue coloring and is undoubtedly the most celebrated diamond in the world. What is the origin of its intensely prized blue hue?

The Chemical Essence

Throughout history civilization has treasured the rarity and beauty of fancy colored diamonds. The stunning diamond from India known as the "Hope Diamond," once a part of many royal inventories, is now the premier attraction of the Smithsonian Institution. Although the size of the diamond at 45.52 carats has certainly contributed to the public's interest in the gem, the intense blue-violet color of the stone is generally considered to be its most captivating feature. First described in the mid-1600s by the French merchant traveler Jean Baptiste Tavernier as "un beau violet" (a beautiful violet), the gem also acquired the title "Blue Diamond of the Crown" or the "Royal French Blue" when in possession of King Louis XIV of France. The blue color is attributed to trace amounts of boron in the carbon matrix of the stone. Substitution of carbon atoms by nitrogen leads to yellow diamonds, such as the famous canary yellow 128.51-carat "Tiffany" diamond.

THE CHEMICAL SPECIFICS

The chemical composition and crystal structure of a mineral determine its physical and optical properties. The diamond crystalline lattice structure (Fig. 6) is composed of two interpenetrating face-centered cubic (fcc) lattices. These lattices are characterized by equal lengths of all sides of the lattice (hence a "cubic" structure) and by atoms positioned at each corner of the cubic lattice and at the center of each face of the lattice. The two interpenetrating lattices are displaced by one-fourth of the distance of the lattice edge. Each carbon atom in one fcc lattice is tetrahedrally coordinated with four other carbon atoms in the other fcc lattice. Other materials with the diamond crystal structure include silicon and germanium. Carbon's neighbors in the periodic table, boron and nitrogen, have similar atomic sizes to carbon and can substitute for carbon in the diamond structure. Even a substitution in the parts per million range is sufficient to change the optical properties of diamond from a colorless gem to a pale blue color. Increasing substitution of boron deepens the blue hue.

KEY TERMS

face-centered cubic lattice

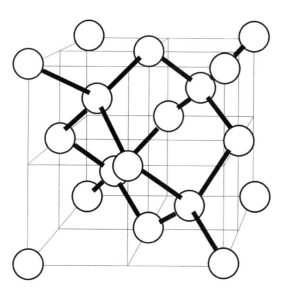

Figure 6 The diamond crystalline lattice structure composed of two interpenetrating face-centered cubic lattices.

RELATED WEB SITES

"Famous Fancy Diamonds: A Brief History," *Gemmology Canada*—Special Issue, Linda Crane, http://www.cigem.ca/423.html

"Gems and Precious Stones: Color in Minerals," Jill Banfield, The Department of Geology and Geophysics, University of Wisconsin-Madison, http://www.geology.wisc.edu/~jill/Lect7.html

"Introduction to Fancy Color Diamonds," Florida Jewelry Appraisers, http://www.colored-diamonds.com/page1.html

"The Nature of Diamonds," American Museum of Natural History, http://www.amnh.org/Exhibition/index.html

"Smithsonian FAQ's: Hope Diamond," National Museum of Natural History, Smithsonian Institution, http://www.si.edu/resource/faq/nmnh/hope.htm

WHY ARE OPALS AND PEARLS IRIDESCENT?

For centuries the gemstone opal has been admired for its captivating and distinctive play of color. The name of this gemstone is attributed to various origins, including the Latin word opalus *and the Sanskrit word* upala, *both meaning precious stone or jewel, and the Greek term* opallios, *translating as "to see a change of color." Although revered as "the queen of gems" in Shakespeare's* Twelfth Night, *the opal suffered in popularity in the 19th century when Sir Walter Scott associated the opal with bad luck in the novel* Anne of Geierstein. *What is the chemical origin of the opal's dramatic rainbow of flashes of colors?*

THE CHEMICAL ESSENCE

A variety of minerals are prized for their exquisite beauty, rarity, and exceptional durability. These extraordinary materials are classified as gemstones. One such mineral, *silica,* with a chemical composition of SiO_2 (silicon dioxide), exhibits several crystal structures. Several gemstones are crystalline forms of silica, including amethyst, aquamarine, emerald, garnet, peridot, topaz, tourmaline, and zircon.[1]

The gem *opal* is an essentially noncrystalline or amorphous form of silica that contains one or two silica minerals. Opal's gemstone quality is primarily derived from its *iridescence*—the interference of light at the surface or in the interior of a material that produces a series of colors as the angle of incidence changes. A very common example of this phenomenon is the observation of color fringes or bands when looking at a thin layer of spilled gasoline. What is the source of this iridescence in opal? A combination of interference and diffraction effects result from the structure of opal. Opal is an example of a *solid emulsion*—a heterogeneous mixture that is composed of a solid host material and a dispersed liquid. The silica minerals serve as the dispersing medium with liquid water suspended within the solid substance. Submicroscopic spheres of the silica minerals act as a three-dimensional diffraction grating, producing a spectacular array of colors. The water content is variable from 1 to 30%, with typical levels at 4 to 9%.[2] The water content is related to the temperature of the host rock at the time the opal formed.[3] Although opal is fundamentally colorless, impurities often impart color from the yellows, oranges, and reds typical of "fire opal" derived from iron oxides to the very dark gray, blue, or black color of the rare "black opal" arising from manganese oxides and carbon.[4]

Pearls are also considered to be gemstones comprised mainly of calcium carbonate ($CaCO_3$). The luster associated with pearls is a consequence of the solid emulsion nature of pearls, with water again serving as the dispersed substance.

THE CHEMICAL SPECIFICS

All silica minerals have a basic three-dimensional microscopic structure composed of tetrahedral arrangements of oxygen atoms about a central silicon atom, with each oxygen atom shared with a neighboring tetrahedral group. The packing of the tetrahedra vary among the silica minerals, with the most common form of silica—*quartz*—exhibiting a relatively dense packing. Two other forms of silica—*cristobalite* and *tridymite*—have a much more open framework, easily accommodating small amounts of impurities within the three-dimensional structure. The gem opal contains cristobalite, tridymite, or mixtures of these two silica minerals. Because the structure of opal is not crystalline but *amorphous* (lacking in a periodic three-dimensional arrangement of atoms), opal is often considered a *mineraloid* (a naturally occurring "mineral" lacking a crystalline structure). Due to its varying water content, opal is also described as hydrated silicon dioxide or a *hydrate* of silica with a general formula of $SiO_2 \cdot n\, H_2O$. Using a technique known as electron microscopy, the microstructure of precious opal was first revealed in 1963.[5] The iridescent colors associated with opal arise from a semiperiodic arrangement of uniform microspheres of hydrated SiO_2 that are spaced at about the wavelength of visible light. These microspheres range in size from 200 to 300 nanometers in diameter[6] and serve as a diffraction source for refracted and reflected light.

Thermal analysis techniques reveal that water is bound in opal in more than one manner.[7] Most of the water is physically held in inclusions or microscopic pores within the opal, that is, in spaces between the microspheres. Water held in this manner can escape through complex systems of microscopic fissures or cracks, induced by temperatures even below 100°C. Some water is held within the opal via chemical bonding ("adsorption") to the surfaces of the silica microspheres and is retained to temperatures approaching 1000°C.[7] Furthermore, because the microspheres themselves are composed of much smaller silica particles, water is additionally coated on the surfaces of these minute particles. The porous nature of opal and its thermal sensitivity require special care, for dehydration may result in cracking that greatly diminishes the value of this gemstone.

KEY TERMS

amorphous; crystalline; diffraction; hydrate; solid emulsion

REFERENCES

1. "The Silicate Class," Amethyst Galleries, Inc., http:/mineral.galleries.com/minerals/silicate/class.htm
2. "Silica Mineral," *Encyclopedia Brittanica Online,* http:/www.eb.com:180/cgi-bin/g?DocF=macro/5004/28/94.html
3. "The Gemstone, Opal," Amethyst Galleries, Inc., http://mineral.galleries.com/minerals/mineralo/opal/opal.htm
4. "Opal," http://www.eb.com:180/cgi-bin/g?DocF=micro/439/2.html
5. "Opal under the Microscope: The Microstructure," OpalNet, http:/www.opalnet.com.au/story_html/opal_09_07.html
6. "The Mexican Opal," Carmen Ramirez, Gemmology Canada, Canadian Institute of Gemmology, http://www.cigem.ca/415.html#SCRL5
7. "Of What Is Opal Made? The Chemical Composition of Opal," OpalNet, http://www.opalnet.com.au/story_html/opal_09_09.html

RELATED WEB SITES

"Agates and Opals," Lecture 16b of Geology 306, The University of Wisconsin Geology and Geophysics Department, Professor Jill Banfield, http://www.geology.wisc.edu/~jill/Lect16b.html

"Gems and Precious Stones," The University of Wisconsin Geology and Geophysics Department, Professor Jill Banfield, http://www.geology.wisc.edu/~jill/306.html

"Smithsonian Gem and Mineral Collection," http://galaxy.einet.net/images/gems/gems-icons.html

"Opal: Rainbow of the Desert," OpalNet, http://www.opalnet.com.au/story_html/foreword.html

"Pearls—More Than a Gemstone?" *Gemmology Canada*—Special Edition, Shirley Ford-Bouchard, http://www.cigem.ca/417.html

Chapter 10

Connections to Fabrics and Clothing

HOW IS A FABRIC MADE WATER REPELLENT OR WATERPROOF?

Designing effective water-repellent, water-resistant, or waterproof fabrics to provide protection in inclement weather or during certain outdoor activities requires an understanding of the chemical and physical properties of water. Whether you are sailing in wind-driven rain, hiking in a downpour, or sitting on a wet surface, chemistry can keep you dry!

THE CHEMICAL ESSENCE

A variety of finishes are applied to the surfaces of fabrics to impart a resistance to water. The water-repellent finish must be capable of interacting chemically with the fabric to form a film or coating, but the film must not possess a chemical structure favorable to interaction with water. Waxy coatings, silicone, and Scotchgard are some of the chemical finishes used in the fabric industry. (In reality, more than 100 different Scotchgard products with more than 30 formulations have been designed to optimize the performance on a variety of fibers, including polyester, nylon, cotton, rayon, polypropylene, wool, and blends.)[1] DuPont has introduced a Teflon fabric protector to resist water stains on a variety of fabrics including silk. Although most water-repellent finishes are surface treatments, some protective polymers are capable of penetrating the fabric and encapsulating every

169

fiber throughout the fabric. Treatment with such polymers provides increased resistance to water.

Alternatively, recent developments in the textile industry have led to the design of fabrics that repel water due to the structure of the fabric itself rather than a chemical finish applied to the fiber. GORE-TEX fabric is one example of a waterproof material consisting of a Teflon film (actually, a patented Gore-Tex membrane) glued or laminated to a conventional synthetic fabric such as nylon or polyester. The pores of the membrane are quite small—nine *billion* pores exist per square inch! In particular, each pore is 20,000 times smaller than a raindrop but 700 times larger than a molecule of water.[2] As a consequence, a droplet of liquid water is too large to pass through a pore. The surface tension of water—a property of the liquid that arises from the extensive interactions between water molecules in the liquid state ("hydrogen bonding")—keeps the droplet intact, preventing tiny droplets from forming that could pass through the pores. On the other hand, water vapor from perspiration on the skin can pass through the pores because, as a gas, the water vapor molecules do not collect into droplets.

THE CHEMICAL SPECIFICS

The application of water-repellent finishes to fabrics actually involves a chemical reaction between the material and the finish. Cellulose-based fibers such as cotton possess hydroxyl ($-OH$) groups that exist on the surface of fabrics spun and woven from the fiber. The basic structure of cellulose portrayed in Figure 1 reveals three hydroxyl groups per six-membered ring.

As an example of a chemical reaction between cellulose and a water-repellent finish, consider the organosilicon compound trimethylchlorosilane, $(CH_3)_3SiCl$. This compound possesses strong $Si-C$ bonds but in the vapor phase is capable of reacting with $-OH$ groups to form oxygen-silicon linkages and generate HCl:

$$-OH + (CH_3)_3SiCl \ (g) \rightarrow \ -O-Si(CH_3)_3 + HCl \ (g)$$

Briefly exposing a cotton fabric to trimethylchlorosilane vapor replaces the strong water-attracting hydroxyl groups with water-repellent $-O-Si(CH_3)_3$ groups on the fabric surface. The loss of hydrogen bonding capacity is the prime factor for the water-resistant character of the treated fabric. Fluorocarbon-based finishes such as Scotchgard are applied to fibers in much the same fashion. The fluorochemical is composed of two major parts: a water-soluble portion that reacts with a functional group on the fiber (such as the hydroxyl group in cellulose) and an inert portion (due to the strength of carbon-fluorine bonds) that repels water.

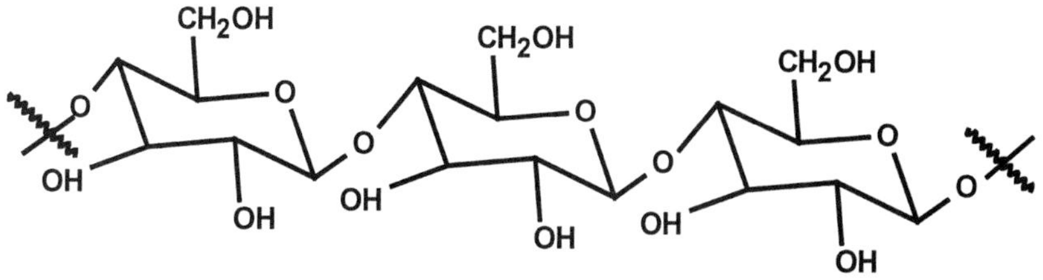

Figure 1 Representation of the repeating monomeric units of the polymer cellulose.

Waterproof GORE-TEX fabric is an innovative application of chemistry that provides an extremely useful consumer material. The technology developed by W. L. Gore and Associates uses unique microstructures to control the porosity of the fabric. The patented GORE-TEX membrane is a composite of two materials—expanded polytetrafluoroethylene (ePTFE) to provide the membrane structure and polyalkylene oxide polyurethane-urea (POPU) to protect the membrane from soiling or contamination. Both the physical and chemical characteristics of the PTFE membrane give rise to the waterproof nature of GORE-TEX fabric. The pore structure of the membrane acts as an impermeable barrier to liquid water whereas the hydrophobic ("water-hating") nature of the polymeric PTFE repels water. The nonpolar carbon-fluorine bonds of the PTFE membrane do not permit hydrogen bonding between water and the fluorine atom. Liquid water molecules retain an affinity for each other rather than for the PTFE membrane, thereby maintaining water droplet size. As the membrane structure is integral to the functioning of GORE-TEX fabric, contamination or blockage of the pores by body oils, soaps, lotions, cosmetics, insect repellants, and dry cleaning solvents must be avoided. Saturation of the membrane with the oleophobic ("oil-hating") substance POPU ensures that contamination does not occur. Lamination of the GORE-TEX membrane to a conventional fabric adds durability.

KEY TERMS

hydrogen bonding; hydrophobic

REFERENCES

1. "3M Scotchgard Fabric Protector," http://www.linx3.com/immaculate/scotchgard/scotchga.htm

2. "GORE-TEX OUTERWEAR," The W. L. Gore & Associates, Inc., http://www. gorefabrics.com/allweather/outerwea.htm

RELATED WEB SITES

"Innovative Lives: Patsy Sherman and the invention of Scotchgard," Smithsonian Institution, http://www.si.edu/lemelson/centerpieces/ilives/lecture11.html
"No More Crying Over Spilt Milk," DuPont On-Line Magazine, October 6, 1996, http:// www.dupont.com/corp/products/dupontmag/sepoct96/milk.html
3M Innovation Network—The Homepage of the 3M Company, http://www.mmm.com
"What Is Scotchgard protection," 3M Technology, 3M Australia, http://www.3mco.fi/intl/AU/ technology/sgard.html
The W. L. Gore & Associates, Inc. homepage, http://www.gorefabrics.com

WHY DOES A BULLET-PROOF VEST WORK? WHAT IS THE COMPOSITION OF A BULLET-PROOF VEST?

Security personnel often rely on the integrity of bullet-proof vests to protect their lives. The innovative chemistry of the fabric creates the incredible strength of these ballistic barriers.

THE CHEMICAL ESSENCE

Bullet-resistant apparel are often constructed of manufactured fibers known as *aramids*. Kevlar, developed by DuPont in 1971, is one example of this class of fibers. The unique properties exhibited by Kevlar—ultrahigh strength and stiffness—arise from its chemical composition. How can chemical identity impart such features as structural rigidity, cut resistance, flame resistance, even inertness to chemical attack? Kevlar is an example of a synthetic polymer composed of long, rodlike chains of stiff chemical units ("monomers") that are strongly linked ("bonded"). An extensive network of interactions between the linear chains imparts further rigidity and stability. This combination of intermolecular (i.e., between two or more molecules) interactions and intramolecular (i.e., within the same molecule) bonds gives Kevlar a strength per weight that is five times that of steel.

In addition to ballistic body armor, there are numerous applications for the strength of fibers like Kevlar. As a durable yet lightweight and flexible fiber, Kevlar is used in the manufacture of gloves and sleeves that offer protection from heat and cuts. Protective clothing for firefighting and for sports, such as bicycle helmets, batting gloves, and NHL goalies' face masks, are additional apparel uses for Kevlar. Sporting equipment including canoes and kayaks, kites, golfclubs, ski poles, and in-line skates use Kevlar as a composite fiber to increase durability, add stiffness, and reduce weight. As a reinforcement fiber, Kevlar also offers high performance for puncture-resistant tires for demanding terrain and for wear-resistant automotive brake pads and hoses. Airships constructed from Kevlar are sealed on the inside with a film such as saran or mylar.

On the horizon is a new technology for ballistic protective clothing using nonwoven fabrics. Because the weaving process can be detrimental to the linear alignment that gives fibers such as Kevlar their strength, the scientists at AlliedSignal have designed a process that maintains the fiber orientation. Both aramid fibers and polyethylene-based fibers such as Spectra fiber de-signed by the chemists at AlliedSignal may be employed in this technology. How is this nonwoven fabric created? Parallel strands of the linear synthetic fiber are mounted side by side with a flexible *thermoplastic resin* (i.e., a polymer that attains the characteristics of plastic upon heating). Two such layers of unidirectional fibers are arranged at 90° to cross the fibers at right angles and then are fused with heat and pressure to create a composite structure. Two sheets of thin, flexible polyethylene film are then sandwiched on either side of the composite structure to reduce contact with dirt, moisture, and abrasives. These laminated rolls of fabric are ultralight and per weight ten times stronger than steel. Fabricated into durable composite structures, Spectra-based prod-ucts are providing increased effectiveness as ballistic barriers.

THE CHEMICAL SPECIFICS

Aramids are polymeric species that contain both aromatic and amide groups. The aromatic functionality consists of a substituted benzene ring, whereas the amide moiety is a $C{=}O$ carbonyl group bonded to a nitrogen atom with at least one hydrogen substituent, $C({=}O){-}NHR$. The combina-tion of the distinct chemical properties of these two functionalities provide the basis for the ultrahigh strength of the polymer. The aromatic portion of the polymer confers strength within a single polymer chain by limiting bond rotation, whereas the amide linkage is essential for strong hydrogen bonding between polymer chains.

Kevlar, the trademark for poly-*para*-phenylene terephthalamide, is an example of an aramid polymer formed from two different monomeric units, a diamine (i.e., a monomer with two amine or $-NH_2$ groups) and a dicar-boxylic acid (i.e., a monomer with two $-COOH$ or carboxyl groups). The *condensation reaction* of terephthalic acid and 1,6-hexamethlenediamine produces the amide linkage and a water molecule as byproduct, as shown in Figure 2. Mixing a stoichiometric amount of diamine and dicarboxylic

Figure 2 The condensation reaction of terephthalic acid and 1,6-hexamethlenediamine to produce the amide linkage in Kevlar and a water molecule as by-product.

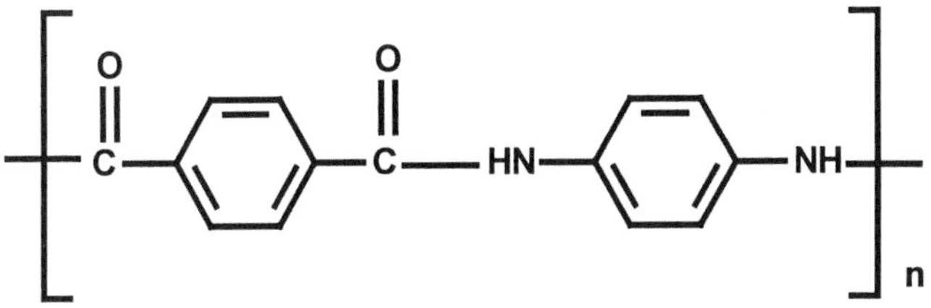

Figure 3 The repeating unit of the polymer Kevlar.

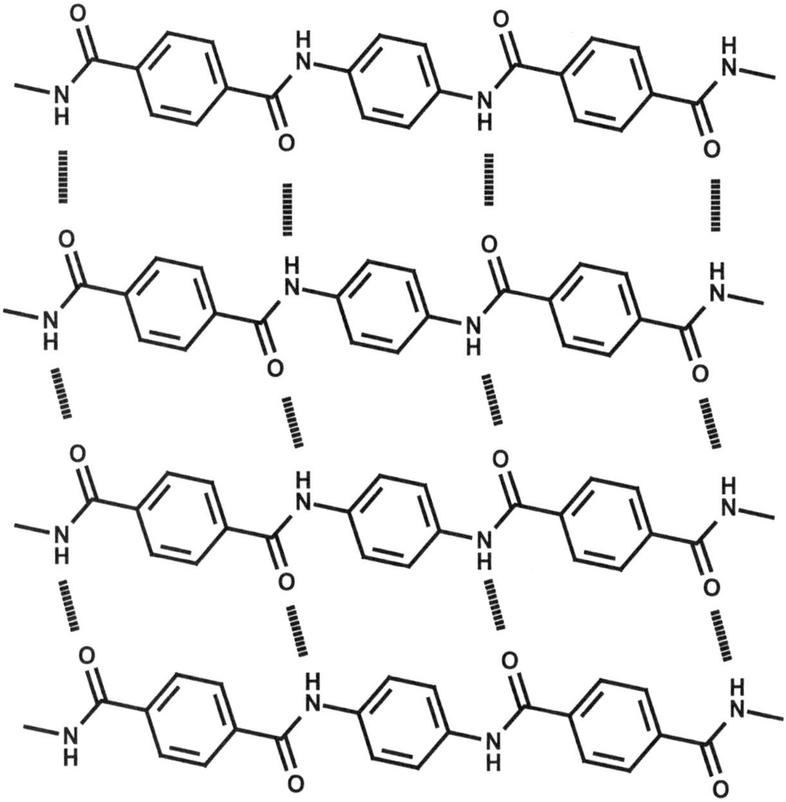

Figure 4 Strong interchain hydrogen bonding in Kevlar.

acid (i.e., a 1:1 ratio) yields a linear polymer with the repeating unit shown in Figure 3. Why is this linearity important to the functioning of the polymer? The rodlike nature of the polymer orients the molecules in solution in essentially a single direction, facilitating the strong interchain hydrogen bonding needed to create the cross-linked matrix (Fig. 4).

The first synthesis of Kevlar by solution polymerization was reported by S. L. Kwolek, P. W. Morgan, and W. R. Gorenson of DuPont in U.S. Patent No. 3,063,966 (1962). In 1980, Stephanie Kwolek won the American Chemical Society's Award for Creative Invention, and on July 22, 1995, she was inducted into the National Inventors Hall of Fame in Akron, Ohio. In 1996 Stephanie Kwolek was awarded the National Medal of Technology "for her contributions to the discovery, development and liquid crystal processing of high-performance aramid fibers which provide new products worldwide to save lives and benefit humankind".[1]

KEY TERMS

aramid; amide bond or amide linkage; condensation polymer or condensation reaction; hydrogen bond

REFERENCE

1. "National Medal of Technology; 1985–1997 Recipients," http://www.ta.doc.gov/medal/Recipients.htm

RELATED WEB SITES

"Batter Up," DuPont On-Line Magazine, August 3, 1997, http://www.dupont.com/corp/products/dupontmag/97/wearforce.html
"A Big Hand for Kevlar Plus," DuPont On-Line Magazine, July 6, 1997, http://www.dupont.com/corp/products/dupontmag/97/kevlarplus.html
DuPont Kevlar Home Page, http://www.dupont.com/afs
First Defense International Home Page, http://www.firstdefense.com/html/bulletproof_vests.htm
"Innovative Lives: Stephanie Kwolek and the invention of Kevlar," Smithsonian Institution, http://www.si.edu/lemelson/centerpieces/ilives/lecture05.html
"Kevlar: The Wonder Material," Microworlds: Exploring the Structure of Materials, Lawrence Berkeley Laboratory, http://www.lbl.gov/MicroWorlds/Kevlar/KevlarIntro.html
"Out of the Flying Plane, Into the Fire," DuPont Magazine Online, January 25, 1998, http://www.dupont.com/corp/products/dupontmag/98/forestfire.html
"Respect and the GT," DuPont Magazine Online, February 22, 1998, http://www.dupont.com/corp/products/dupontmag/98/Article8.html

"Stephanie Louise Kwolek," Inventor of the Week Archives: The Lemelson-MIT Prize Program, http://web.mit.edu/invent/www/inventorsI-Q/kwolek.html

WHY IS A COTTON TOWEL MORE EFFECTIVE IN CLEANING UP WATER SPILLS THAN A TOWEL MADE FROM A SYNTHETIC POLYESTER FIBER? WHY DO COTTON FABRICS TAKE LONGER TO DRY THAN SYNTHETICS?

You've no doubt noticed many differences between natural fibers like cotton and synthetic fibers like polyester. Cotton is the most commonly used fiber by designers and manufacturers. U.S. cotton producers and importers of cotton goods into the United States have emphasized this point in their advertisements for cotton, "The Fabric of Our Lives".[1] Nevertheless, the desirable characteristics of polyester have made it the most used and most blended synthetic fiber. The chemical structures of these fibers provide an explanation for the differences in the response of these materials to water.

The Chemical Essence

The effectiveness of a towel in absorbing a water spill depends on the chemical structure of the fiber. One of the governing principles in chemistry is that, for any two substances, the closer their chemical structures, the greater the likelihood of the materials interacting. The attractive interactions on the molecular level are called *intermolecular forces*. Water is a highly *polar* substance with an affinity for other polar materials. Polarity is enhanced when the atoms, bonds, and shape of a molecule lead to an unequal distribution of charge. In addition, a water molecule has the ability to form strong associations with other water molecules through interactions known as "hydrogen bonds". Cotton fibers are also capable of linking with each other or with water through hydrogen bonds. These compatible associations lead to high water absorbency by cotton towels. In contrast, the chemical structure of synthetic polyester fibers does not favor as strong a connection on the molecular level with water, leading to reduced absorbency. The same principles explain why lightweight synthetic fabrics dry much faster than cotton towels and natural fibers.

The Chemical Specifics

The fiber cotton is composed of the polymer cellulose whose structure appears in Figure 5. All of the factors required for *intermolecular* (i.e.,

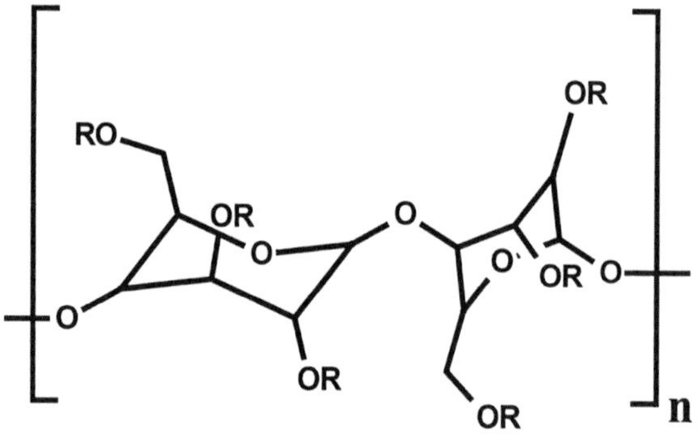

$R = CH_2CH_3$

Figure 5   The repeating unit of the polymer cellulose, the constituent of the fiber cotton.

between two molecules) *hydrogen bonding* are present in cellulose. In particular, recall that a hydrogen bond consists of an arrangement of three atoms, denoted X—H···Y, where the symbol ··· represents the hydrogen bond and X and H are atoms covalently bonded to one another. X is typically the element F, O, or N, that is, a small electronegative element to create a polar bond with the element H. For intermolecular hydrogen bonding the atom Y must be an electronegative element with an unshared pair of electrons and be contained within a second molecule either identical to or distinct from the molecule containing the X—H fragment. The strong polar nature of the X—H bond (due to the differences in electronegativities of X and H) results in a partial positive charge on the hydrogen. This partial charge separation may be denoted as $X^{\delta-}$—$H^{\delta+}$. The slight positive charge on H attracts the lone pair of electrons on Y, creating the so-called *hydrogen bond.* Colinearity of X, H, and Y yields the strongest possible hydrogen bond.

The structure of cellulose reveals the two key components for hydrogen bonding: OH functional groups (the X—H unit) and additional oxygen atoms (the Y atom). Hence, water molecules may form intermolecular

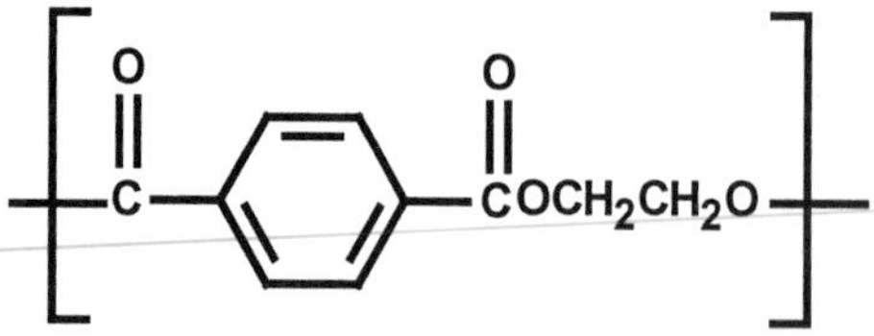

Figure 6   The repeating unit of the condensation polymer Dacron, an example of a polyester.

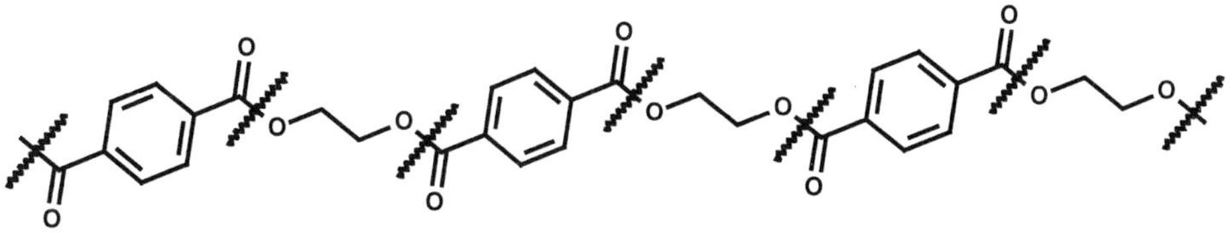

Figure 7 Several linked monomers of the polymer Dacron.

hydrogen bonds with cellulose in two ways: (1) the central oxygen atom in water (acting as Y) can hydrogen bond the hydrogens in the O—H functional groups of cellulose (the X—H unit) or (2) the hydrogen atoms in water (covalently bound to oxygen and thus composing an X—H fragment) can hydrogen bond to the oxygen atoms in cellulose (serving as Y). Thus, extensive intermolecular hydrogen bonding is possible between cellulose and water, enhancing the water capacity of a cotton towel.

As an example of a polyester fiber, consider the condensation polymer Dacron (also sold as a film—Mylar). The monomeric repeating unit of Dacron is shown in Figure 6, with several linked monomers indicated in Figure 7. Although the polyester fiber contains the electronegative oxygen atom (Y) to hydrogen bond to the hydrogen atoms in water (X—H), there are no corresponding X—H units in the polyester to hydrogen bond to the oxygen atom in water (Y). Hence, the extent of hydrogen bonding between water and polyester is significantly reduced from the degree of hydrogen bonding possible between water and cellulose. This explains the relative effectiveness of the two materials in cleaning up water spills.

KEY TERMS

intermolecular forces; hydrogen bonding; polymer

REFERENCE

1. "Cotton Incorporated," http://www.cottoninc.com

WHAT IS AN OPTICAL BRIGHTENER?

"Whiter whites! Brighter brights!" You've heard such claims made by many laundry detergents. The chemical structure of the additives called "opti-

*cal brighteners" provides the essential factors that make these superior deter-
gents possible.*

The Chemical Essence

Natural fibers such as wool and silk and cellulose-containing paper often have a yellowish tinge. To appear yellow these materials must absorb the complementary color—a violet to blue hue (i.e., light of wavelength near 400 nm). (The complementary color is the color on the opposite side of a color wheel.) Natural pigments in these fabrics and in cellulose are responsible for the absorption. Whitening of fabrics and paper can be accomplished using bleaches, but this treatment often degrades the material. Alternatively, an *optical brightener* may be added to replace the blue-violet light that is lost (i.e., not transmitted to our eyes but absorbed by the fiber). An optical brightener is a compound that accomplishes this function by absorbing ultraviolet light and subsequently emitting ("fluorescing") blue visible light. The emitted blue light from the optical brightener replaces the blue light that is absorbed by the fabric, thereby creating a "complete" white light that contains all of the frequencies of the color spectrum. The further "brightening" action of an optical brightener arises when a slight excess of the fluorescent compound is used to convert even more ultraviolet light into visible light. The amount of optical brightener used should be carefully controlled, for an excessive level of optical brightener can cause the fabric to have a blue cast due to the blue emission of light.

The Chemical Specifics

Although many commercial optical brighteners are trade secrets, most of these fluorescent compounds contain one or more ring systems and are derivatives of stilbene (Fig. 8) coumarin (Fig. 9), imidazole (Fig. 10), triazole (Fig. 11), oxazole (Fig. 12), and biphenyl (Fig. 13). The uninterrupted chain

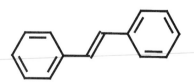

Figure 8 The chemical structure of stilbene.

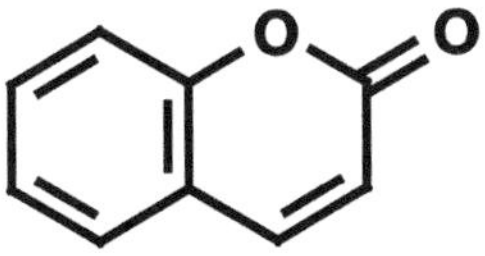

Figure 9 The chemical structure of coumarin.

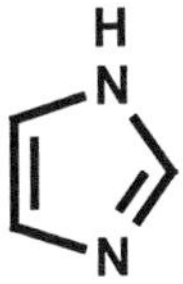

Figure 10 The chemical structure of imidazole.

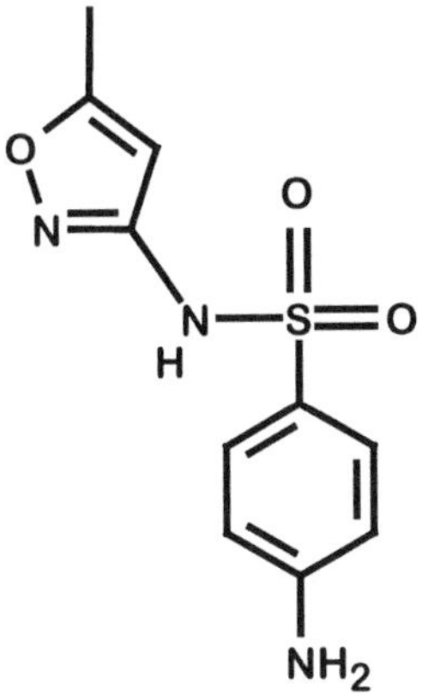

Figure 11 The chemical structure of triazole.

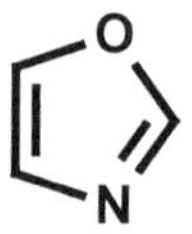

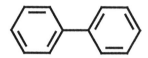

Figure 12   The chemical structure of oxazole.

![biphenyl structure]

Figure 13   The chemical structure of biphenyl.

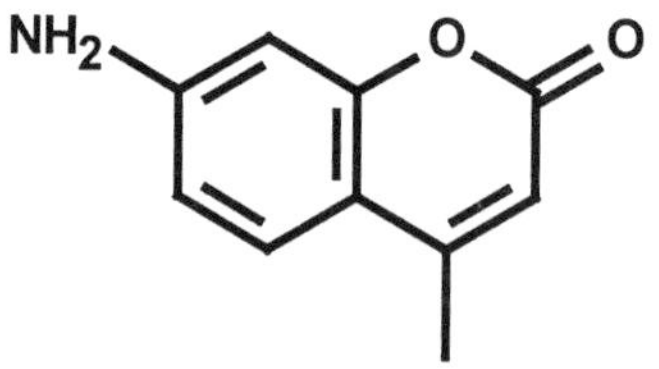

![7-amino-4-methylcoumarin structure]

Figure 14   The chemical structure of 7-amino-4-methylcoumarin.

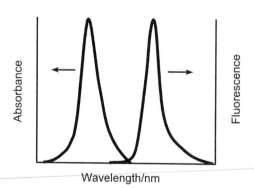

Figure 15   A schematic representation of the normalized absorption and emission spectra of 7-amino-4-methylcoumarin.

of alternating single and double bonds ("conjugated double bonds") in these substances contributes to the absorption of these compounds in the ultraviolet region and their fluorescence in the blue wavelengths of light. One such compound, 7-amino-4-methylcoumarin (Fig. 14), absorbs near 350 nm and emits at 430 nm. These optical parameters are evident in the normalized absorption and emission spectra[1] of this compound that appear schematically in Figure 15. An absorption spectrum is a record of how much light of a particular color is absorbed by a substance as a function of color (i.e., wavelength). An emission (or fluorescence) spectrum, a similar concept, measures the intensity of light of a given color that is emitted by a substance as a function of color (i.e., wavelength). The wavelength of maximum absorption (~350 nm) is in the ultraviolet region of the electromagnetic spectrum. When light principally of this wavelength is absorbed, blue light with a maximum wavelength near 430 nm is emitted.

KEY TERMS

ultraviolet; absorption; emission; fluorescence

REFERENCE

1. "Spectra: 7-amino-4-methylcoumarin," Molecular Probes, Inc., http://www.probes.com/servlets/spectra?fileid=191ph7

RELATED WEB SITES

"The Color Wheel," Painter On-Line, The Maryland Institute, College of Art, http://www.mica.edu/painter/on_line/colorwhe.htm
"Elements of Design: Color Wheel," Susan Gonzalez, Southern Utah University, http://www.iron.k12.ut.us/schools/chs/art/color_wheel.html
"Structure and Colour in Dyes," Bryan D. Llewellyn, http://members.pgonline.com/~bryand/dyes/dyecolor.htm

WHAT PUTS THE "BLUE" IN BLUE JEANS?

"In recognition of his services in the advancement of organic chemistry and the chemical industry, through his work on organic dyes and hydroaromatic compounds."[1]

So reads the announcement by The Royal Swedish Academy of Sciences of the awarding of the 1905 Nobel Prize in chemistry to the German chemist Johann Friedrich Wilhelm Adolf von Baeyer. While we think of "blue jeans" as the quintessential American item of clothing, it was the contributions of the German chemist von Baeyer that put the "blue" in blue jeans and enabled this image of American culture to flourish.

THE CHEMICAL ESSENCE

The dye indigo is responsible for the recognizable blue color of blue jeans. This dye is both a natural dye as well as a synthetic one. Indigo plants, any shrub or herb of the genus *Indigofera,* were once cultivated for the purpose of extracting the indigo pigment. However, the research chemist Johann Friedrich Wilhelm Adolf von Baeyer synthesized indigo in 1880 and formulated its structure in 1883, an accomplishment for which he won the Nobel Prize in chemistry in 1905. The industrial process to synthesize indigo was developed by the German firm Badische Anilin- & Soda-Fabrik, now known as BASF, placing indigo on the market in 1897. The current major producers of indigo dyestuff are Ciba Geigy, ICI, BASF, and Mitsui Touatsu in Japan.

As a synthetic fiber dye, indigo possesses several characteristic that confer the well-known features of blue jeans. The dye is quite colorfast to water and light, although it continually fades in intensity over time. Furthermore, the dye's structure prevents its complete penetration into fibers, yielding the irregular and individual appearance generally valued in jeans. You have probably noticed the white specks of fiber when you look closely at denim material.

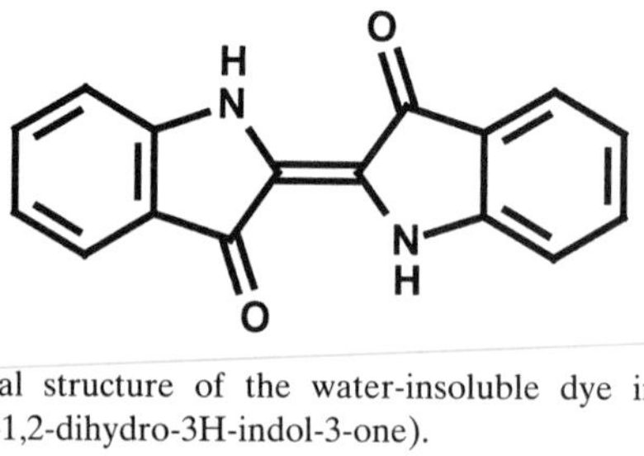

Figure 16 The chemical structure of the water-insoluble dye indigo (2-(1,3-dihydro-3-oxo-2H-indol-2-ylidene)-1,2-dihydro-3H-indol-3-one).

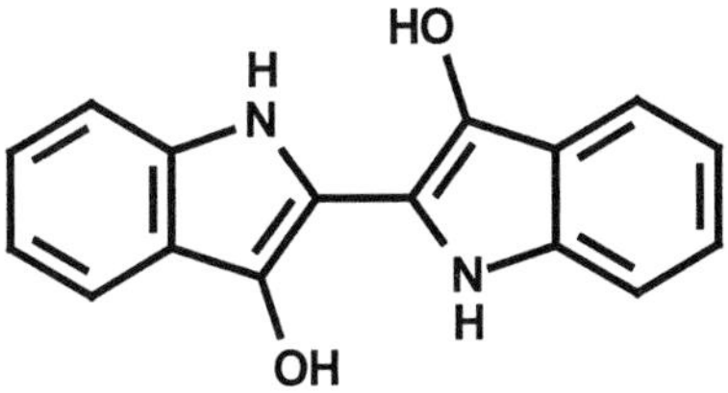

Figure 17 The soluble reduced form of the dye indigo—yellow leucoindigo or [2,2'-biindole]-3,3'-diol.

THE CHEMICAL SPECIFICS

Indigo (or 2-(1,3-dihydro-3-oxo-2H-indol-2-ylidene)-1,2-dihydro-3H-indol-3-one) (Fig. 16) is an example of a water-insoluble dye, contributing to its resistance to fading in water and light. However, to dye denim, a cotton fabric, favorable interactions of the dye and fiber are needed. Cotton is composed of cellulose, a linear polymer with hydroxyl substituents that favor interaction with dyes via hydrogen bonding. To promote interaction with cotton, indigo is applied to textile fibers in its soluble reduced form—yellow leucoindigo or [2,2'-biindole]-3,3'-diol (Fig. 17). The added hydroxyl (−OH) groups in leucoindigo promote water solubility. Sodium hydroxide (NaOH) and sodium hydrosulfite ($Na_2S_2O_4$) are capable of reducing indigo to its colorless form. "Vat dye" is the name given to the class of colored but water-insoluble dyes that must be applied in their reduced state. After dyeing the cotton fiber, the blue color is produced by reoxidizing the indigo. Exposure to air, actually atmospheric oxygen, is sufficient to oxidize the dye to its blue form. Alternatively, chromic acid (potassium dichromate and sulfuric acid) may be used as the oxidizing agent. When laundering jeans, a thimble full of vinegar added to the wash is recommended to keep the jeans dark. Vinegar (or acetic acid) acts as an acid to counterbalance or neutralize the alkaline nature of the detergent in the wash.

KEY TERMS

oxidation; reduction; hydrogen bonding

REFERENCE

1. "Nobel Prize in Chemistry 1905," The Nobel Foundation, http://www.nobel.se/laureates/chemistry-1905-press.html

RELATED WEB SITES

"Indigo Ring Dyeing: Chemistry of Indigo Dyeing," Virkler Company, Charlotte, NC, http://www.virkler.com/expertise/html/chemistry.html

Chapter 11

Connections to Personal Care

WHY DO COSMETIC COLD CREAMS FEEL COOL WHEN APPLIED TO THE SKIN?

The terminology "cold cream" aptly describes the soothing and cooling sensation of these cosmetics. Some basic chemistry fundamentals explain this highly marketed phenomenon.

THE CHEMICAL ESSENCE

The cooling sensation experienced after applying a cosmetic cold cream on one's skin is the result of the evaporation of an alcohol (e.g., ethanol) contained in the cold cream. Formulators of skin products include ethanol to achieve a variety of benefits. For example, alcohol enhances the ability of the components in the cold cream to dissolve. For the consumer, the presence of alcohol eases the application of the cream on the skin, enhances the perfume quality of the mixture, and provides a cooling effect on the skin.

Why does the evaporation of an alcohol achieve a cooling effect? Evaporation is a process that requires heat (an *endothermic* process). The skin in the area of application provides the necessary amount of heat needed to evaporate the alcohol, leading to the localized cooling sensation.

THE CHEMICAL SPECIFICS

Although the particular ingredients of personal care products are often proprietary, one manufacturer, ICI Surfactants, has provided the formulations for two of its "hydroalcoholic skin care emulsions".[1] In each of these products, ethanol is present at a level of 20% by weight. Chesebrough-Pond's, Inc., in the patent abstract for a cold cream cosmetic composition[2], indicates that a *polyhydric alcohol* (an alcohol with more than one hydroxyl group, $-OH$) with two to six carbons is the volatile species that provides the cooling sensation.

The relatively low boiling points and correspondingly high vapor pressures of substances such as ethanol and polyhydric alcohols are factors in promoting the ease with which these materials evaporate and provide a cooling sensation. For example, the vapor pressure of ethanol is 59 torr at 298 K (25°C).[3] How does this value compare with that for water, a substance that we recognize also evaporates easily from one's skin? Water has a vapor pressure of 24 torr at the same temperature:[4] thus ethanol vaporizes even much more readily than water. (Recall that the higher the vapor pressure of a liquid at a fixed temperature, the greater the tendency to escape to the gas phase.)

KEY TERMS

vapor pressure; volatility; evaporation; vaporization; endothermic

REFERENCES

1. "Hydroalcoholic Skin Care Emulsions," ICI Surfactants Skin Care Products, http://www.surfactants.com/HydroAlcoholic.html
2. "Cosmetic Compositions including Polyisobutene," U.S. Patent 5695772, issued December 9, 1997, http://sciweb.com/patents/patents_424_12_31_97.html
3. Lide, D. T. (1993). Properties of common solvents. *In: Handbook of chemistry and physics,* 74th ed., p. 15-46. CRC Press, Boca Raton, FL.
4. Lide, D. T. (1993). Vapor pressure of water from 0 to 370°. *In: Handbook of chemistry and physics,* 74th ed., p. 6-15. CRC Press, Boca Raton, FL.

WHAT IS AN ALPHA HYDROXY ACID?

Many major cosmetic companies have marketed "anti-aging" creams and skin-renewal products containing alpha hydroxy acids as the miracle ingredi-

ents. These particular components are easily identified in a list of a lotion's ingredients, provided you know some simple chemistry nomenclature.

THE CHEMICAL ESSENCE

Many antiwrinkle creams and lotions containing alpha hydroxy acids claim to reverse the effect of sun damage and aging on skin. Most of these products improve the skin's appearance and texture by accelerating the natural process by which the skin replaces its aging outer layer (the epidermis) with new cells. By causing the skin to peel, alpha hydroxy acids temporarily enhance the skin's appearance by revealing "healthier-looking" skin in the lower layers. The chemical identity of the antiaging ingredient, its concentration, its pH or acidity, and the presence of other ingredients in the skin product determine the effectiveness of the acid as an exfoliator.[1]

Alpha hydroxy acids (AHAs) are water soluble substances and thereby penetrate the outermost epidermal skin layers.[2] In contrast, beta hydroxy acids (BHAs) are lipid (fat) soluble and are capable of penetrating to the underlying layers of skin (the dermis) located 1 to 5 mm below the surface of the skin.[3] Most AHAs are derived from plant materials and marine sources. Commonly used AHAs include malic acid (found in apples), ascorbic acid (a common ingredient in numerous fruits), glycolic acid (a constituent of sugar cane), lactic acid (a component of milk), citric acid (naturally abundant in citrus fruits), and tartatic acid (found in red wine). A common BHA is salicylic acid (an ingredient in aspirin).

THE CHEMICAL SPECIFICS

An alpha hydroxy acid is an organic carboxylic acid in which an additional hydroxyl functional group ($-OH$) is present at the alpha position (i.e., on the carbon adjacent to the carboxyl functionality) $-COOH$. Figure 1

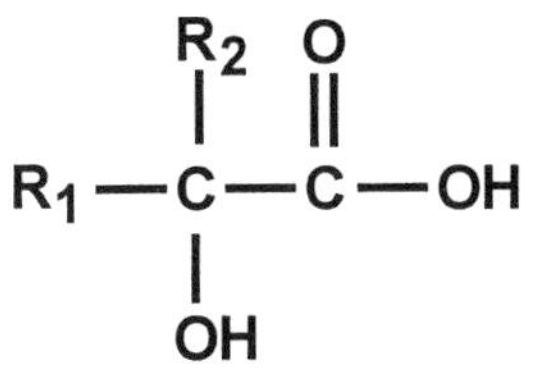

$$R_1 - \underset{\underset{OH}{|}}{\overset{\overset{R_2}{|}}{C}} - \overset{\overset{O}{\|}}{C} - OH$$

Figure 1 The structure of an alpha hydroxy acid, an organic carboxylic acid in which an additional hydroxyl functional group ($-OH$) is present at the alpha position, that is, on the carbon adjacent to the carboxyl functionality, $-COOH$.

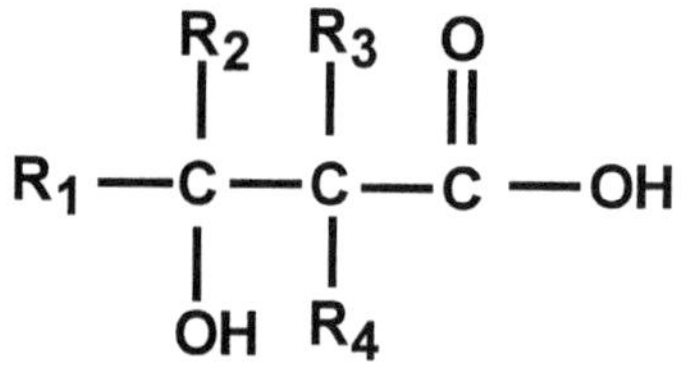

Figure 2   The structure of a beta hydroxy acid, a substituted carboxylic acid in which the hydroxyl substituent is attached to the beta position, that is, at a carbon atom two positions away from the carboxylic acid functionality.

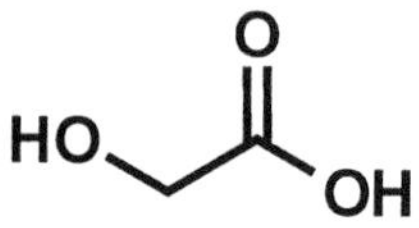

Figure 3   The chemical structure of glycolic acid or alpha-hydroxyethanoic acid.

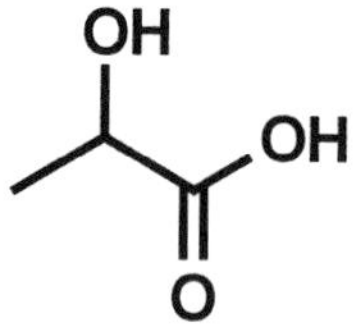

Figure 4   The chemical structure of lactic acid.

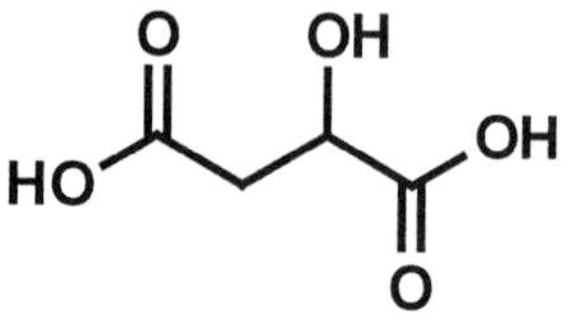

Figure 5   The chemical structure of malic acid.

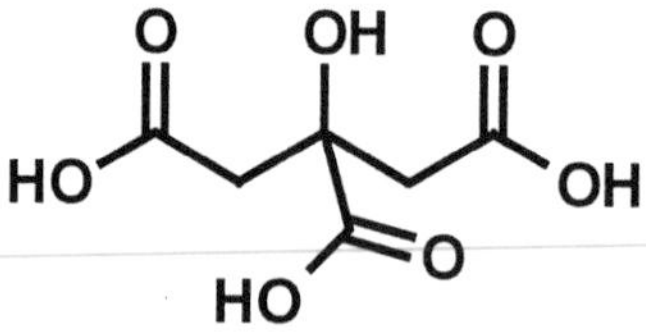

Figure 6   The chemical structure of citric acid.

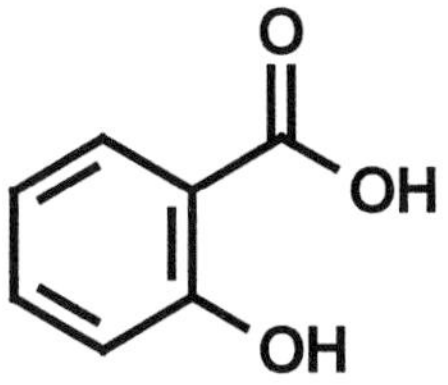

Figure 7 The chemical structure of mandelic acid.

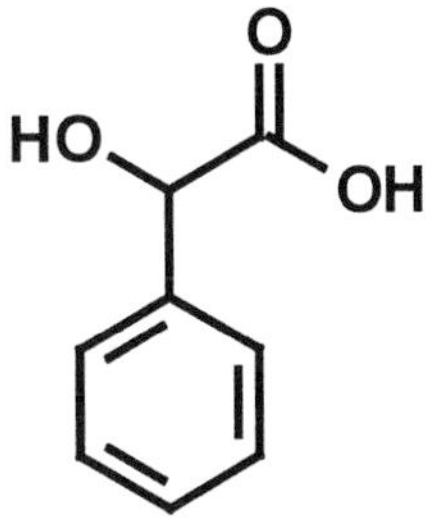

Figure 8 The chemical structure of ascorbic acid.

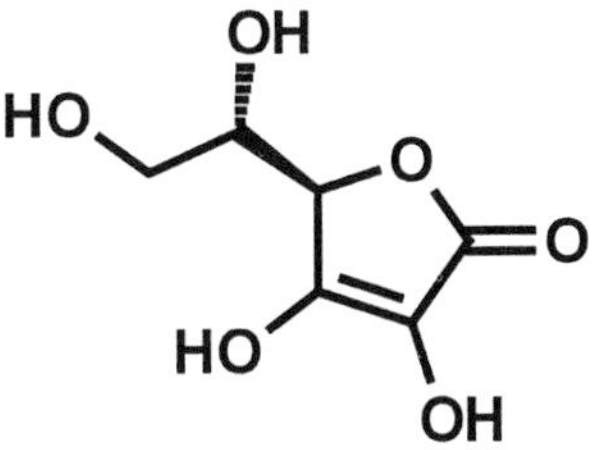

Figure 9 The chemical structure of tartartic acid.

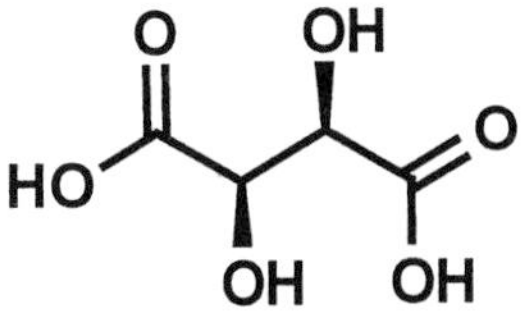

Figure 10 The chemical structure of salicylic acid.

presents the structure of a generic AHA. A beta hydroxy acid is also a substituted carboxylic acid in which the hydroxyl substituent is attached to the beta position (i.e., at a carbon atom two positions away from the carboxylic acid functionality) given by the structure in Figure 2.

Some common alpha hydroxy acids are glycolic acid or alpha-hydroxy-ethanoic acid (Fig. 3), lactic acid (Fig. 4), malic acid (Fig. 5), citric acid (Fig. 6), mandelic acid (Fig. 7), ascorbic acid (Fig. 8.), and tartartic acid (Fig. 9). Salicylic acid (Fig. 10) is the most common beta hydroxy acid, although citric acid could also be classified as a BHA.[3]

KEY TERMS

carboxylic acid; carboxyl functionality; alpha hydroxy acid; beta hydroxy acid

REFERENCES

1. Straight talk about anti-wrinkle products. (1997, October 20). *Oasis,* Mayo Health Clinic, Mayo Foundation for Medical Education and Research, http://www.mayohealth.org/mayo/9710/htm/wrinkles.htm
2. "Alpha Hydroxy Acids," Misty Isle Library, http://www.mistyisle.com/alpha.htm#top
3. "Beta Hydroxy Acids," Misty Isle Library, http://www.mistyisle.com/beta.htm

RELATED WEB SITES

"Alpha Hydroxy Acids in Cosmetics," U.S. Food and Drug Administration, *FDA Back-grounder,* July 3, 1997, http://vm.cfsan.fda.gov/~dms/cos-aha.html
"Alpha Hydroxy Acids for Skin Care: Smooth Sailing or Rough Seas?" Paula Kurtzweil, U.S. Food and Drug Administration, FDA Consumer, March–April 1998, http://vm.cfsan.fda.gov/~dms/fdacaha.html

WHAT IS AN "ALCOHOL-FREE" COSMETIC?

Before you reach for that "alcohol-free" product, stop and consider the features that you expect and desire in your cosmetics. The class of substances known to chemists as alcohols is commonly confused with the consumer interpretation of "alcohol." Knowing the chemist's terminology will help you select the safest and most beneficial formulation.

The Chemical Essence

The United States Food and Drug Administration reports that consumers often inquire about the meaning of the phrase "alcohol-free" when describing a cosmetic product. The term "alcohol-free" specifically applies to the absence of the substance known as *ethanol* or *ethyl alcohol* or *grain alcohol*.[1] An alcohol-free product is often desired by users wishing to avoid drying effects that are often attributed to many alcohol-containing products. Nevertheless, many desirable features of cosmetics—moisturizing, smoothness, emolliency, ease of application, fast drying, and so on, arise from the presence of substances that belong to the broader chemical family of substances known simply as alcohols.

The Chemical Specifics

An alcohol is a member of a class of chemical substances that contain the functional group known as the *hydroxyl* group. This functionality is designated $-O-H$ (i.e., a hydrogen atom bonded to an oxygen atom that is subsequently bonded to another atom) in particular, carbon. Cosmetic products, whether labeled "alcohol free" or not, may contain members of the alcohol family.[2,3] Alcohols confer many highly desirable characteristics to personal care products. For example, many alcohols are key ingredients to act as moisturizers, including cetyl alcohol or hexadecanol (Fig. 11) in facial makeup, hair products, and deodorants; 2-octyl-1-dodecanol (Fig. 12) in skin conditioners; panthenol (Fig. 13) in hair conditioners; stearyl alcohol or octadecanol (Fig. 14); and lanolin alcohols (a mixture of 33 high molecular weight alcohols) in skin and hair conditioners. The alcohol α-tocopherol (vitamin E) (Fig. 15) is used as an antioxidant and preservative to prevent or reduce product deterioration. Phenoxyethanol (Fig. 16) is used in synthetic rose oils and soaps as the rose fragrance component. Many alcohols are included for the solvency properties, that is, their ability to dilute the formulation to the desired consistency. In addition to ethanol (Fig. 17), the alcohols employed for this purpose are substances with more than one hydroxyl group, including glycerin or 1,2,3-propanetriol (Fig. 18), propylene glycol or 1,2-propanediol (Fig. 19), and butylene glycol or 1,4-

Figure 11   The chemical structure of cetyl alcohol or hexadecanol.

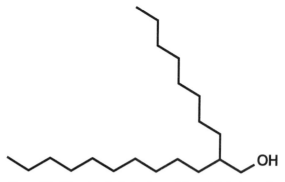

Figure 12   The chemical structure of 2-octyl-1-dodecanol.

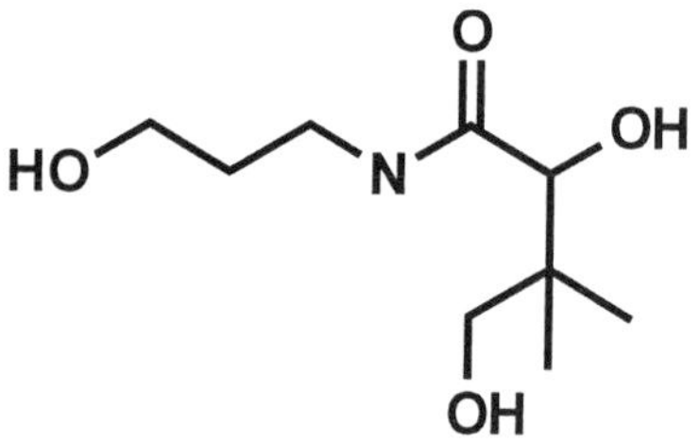

Figure 13   The chemical structure of panthenol.

Figure 14   The chemical structure of stearyl alcohol or octadecanol.

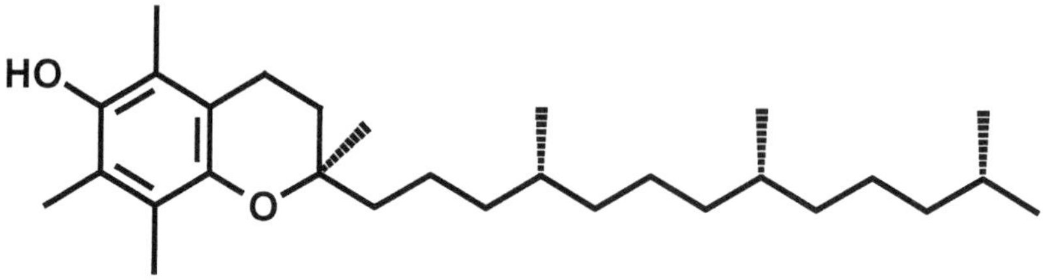

Figure 15   The alcohol α-tocopherol known as vitamin E.

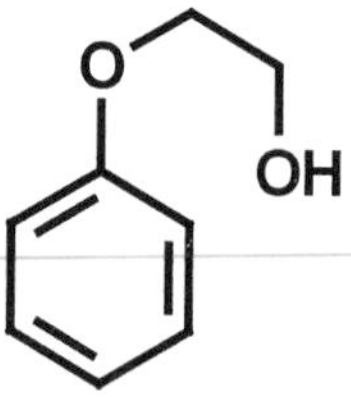

Figure 16   The chemical structure of phenoxyethanol.

Figure 17   The chemical structure of ethanol.

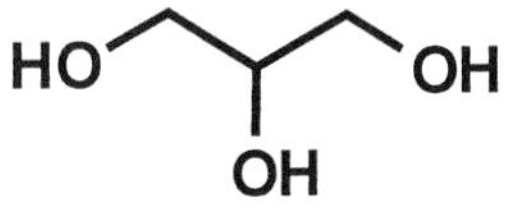

HO~~~OH
OH

Figure 18   The chemical structure of glycerin or 1,2,3-propanetriol.

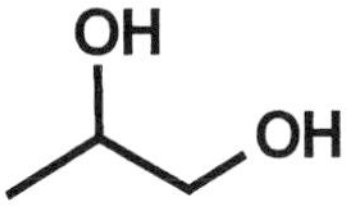

OH
OH

Figure 19   The chemical structure of propylene glycol or 1,2-propanediol.

HO~~~OH

Figure 20   The chemical structure of butylene glycol or 1,4-butanediol.

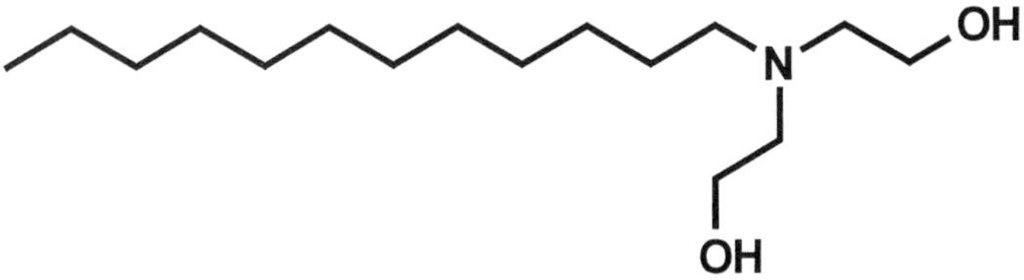

N
OH
OH

Figure 21   The chemical structure of lauric acid diethanolamine or lauramide DEA.

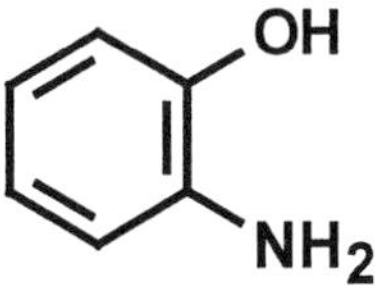

OH
NH₂

Figure 22   The chemical structure of aminophenol or hydroxyaniline.

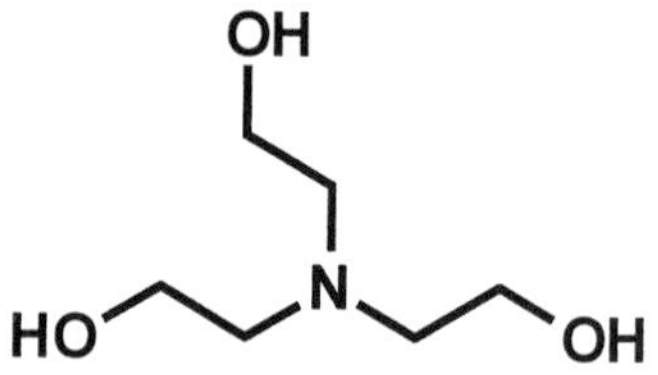

Figure 23 The chemical structure of triethanolamine.

butanediol (Fig. 20). Lauric acid diethanolamine or lauramide DEA (Fig. 21) is often included for its emulsifying properties and foam action. Many permanent hair dye formulations contain derivatives of aminophenol or hydroxyaniline (Fig. 22). Triethanolamine (Fig. 23) is an alcohol included as a pH adjuster in transparent soap. The functions and benefits of alcohols in cosmetic formulations is endless! Users of cosmetic products should probably be quite relieved to learn that their products are not completely hydroxyl free.

KEY TERMS

alcohol; hydroxyl group

REFERENCES

1. "Alcohol Free," U.S. Food and Drug Administration, Center for Food Safety and Applied Nutrition, Office of Cosmetics Fact Sheet, February 23, 1995, http://vm.cfsan.fda.gov/~dms/cos-227.html
2. "Mineral Oil Based Skin Care Emulsions," ICI Surfactants, http://www.surfactant.com/Mineral.html
3. "Chemical Ingredients Found in Cosmetics," U.S. Food and Drug Administration, *FDA Consumer,* May 1994, http://vm.cfsan.fda.gov/~dms/cos-chem.html

WHAT CAUSES THE COOLING SENSATION FOUND IN MANY TOOTHPASTES AND BREATH FRESHENERS?

"Invigorating mint taste for a clean, healthy mouth and fresh breath that lasts." "Clean, cool, refreshing taste." These claims accompany many advertisements for toothpastes and breath fresheners. What is the source of the refreshing flavor and cooling effect that consumers readily associate with these products?

THE CHEMICAL ESSENCE

We often experience a cooling, refreshing sensation when we use various toothpastes and mouthwashes. Manufacturers of such personal hygiene products include one or more key ingredients to provide both a pleasant taste and a bracing feeling of coolness. These components include peppermint oils, spearmint oils, and menthol. These substances belong to a class of materials known as essential oils—the highly concentrated, volatile, aromatic essences of specific plants. To acquire these key substances, the flowers, stems, bark, and leaves of plants must be harvested at exact growth stages, under certain weather conditions, and even at specific times of the day.[1] Specific constituents in these essential oils stimulate the nerves that sense cold and depress those nerves that detect pain. This sensory process is known as *chemoreception* and is initiated when certain chemical stimuli come in contact with chemoreceptors, those specialized cells in the body that convert the immediate effects of such substances directly or indirectly into nerve impulses. The body subsequently responds, producing its own "warming effect" as blood flows into the area of application. This physical sensation impresses the senses as a "medicinal" effect and is partially responsible for peppermint's long history of use as medicine.

THE CHEMICAL SPECIFICS

Peppermint oil is the volatile oil that is extracted from the fresh leaves of the flowering plant of *Mentha piperita* via steam distillation. Peppermint oil contains not less than 44% menthol. American peppermint oil contains from 50 to 78% of free *l*-menthol (Fig. 24) and from 5 to 20% combined in various esters such as menthyl acetate (Fig. 25). The essential oil also contains *d*-menthone, *l*-menthone (Fig. 26), cineole (Fig. 27),

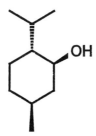

Figure 24 The chemical structure of l-menthol.

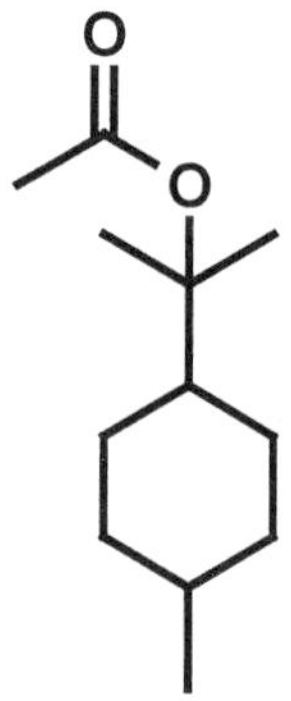

Figure 25   The chemical structure of menthyl acetate.

d-isomenthone, d-neomenthone, and the monoterpene derivative mentho-furan (Fig. 28).[2] The alcohol l-menthol and the ester menthyl acetate are responsible for the pungent and refreshing odor associated with peppermint oil. The specific constituents are found to depend on the age of the leaves, with menthol and menthyl acetate preferably found in older leaves and preferentially formed during long daily sunlight periods. The ketones menthone (Fig. 26) and pulegone (Fig. 29) (and the monoterpene derivative menthofuran, refer to Fig. 28) have a less delightful fragrance and appear to a greater extent in young leaves. The formation of these latter ingredients occurs predominantly during short days. The cooling sensation is associated with the l-menthol optical isomer (i.e., the levorotatory form that rotates the plane of polarized light to the left). Recall that optical isomerism results when two structures are identical in composition and bonding but are nonsuperimposable mirror images of each other. As a consequence of their three-dimensional structure, optical isomers rotate the plane of polarization of a beam of polarized light that is directed through them in different directions. The levorotatory or l form rotates the plane of polarized light to the left; the dextrorotatory or d form rotates the plane of polarized light to the right.

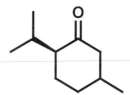

Figure 26   The chemical structure of l-menthone.

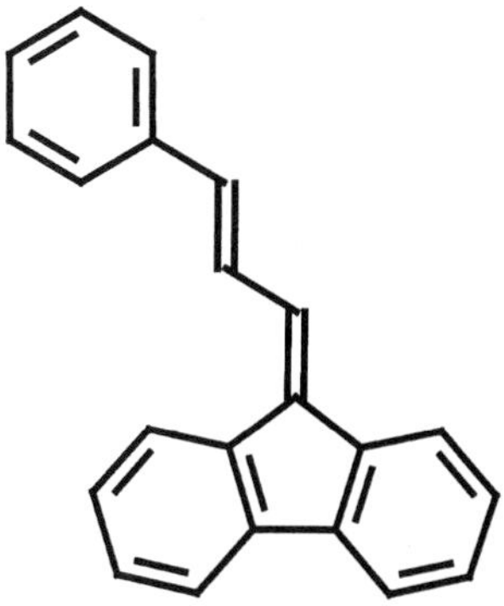

Figure 27 The chemical structure of cineole.

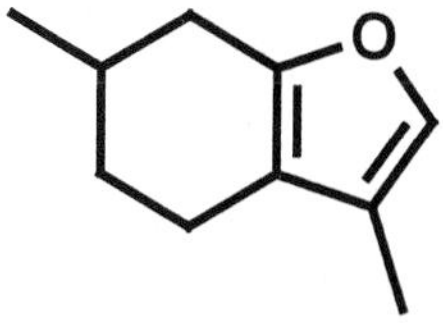

Figure 28 The chemical structure of the monoterpene derivative menthofuran.

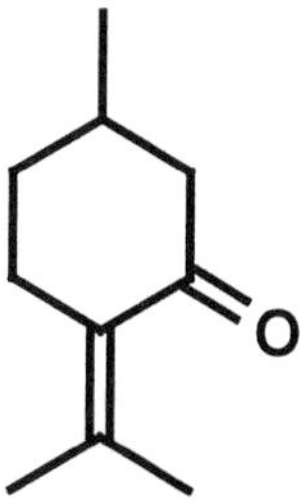

Figure 29 The chemical structure of pulegone.

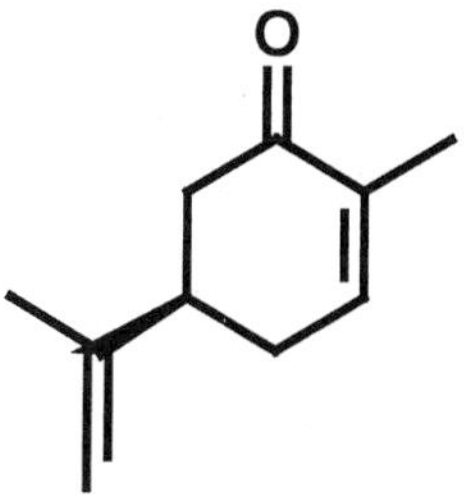

Figure 30 The chemical structure of *l*-carvone.

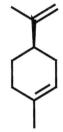

Figure 31   The chemical structure of *l*-limonene.

The main constituents of spearmint oil are *l*-carvone (Fig. 30) and *l*-limonene (Fig. 31). Oil of spearmint contains from 45 to 60% *l*-carvone, 6 to 20% of alcohols, and 4 to 20% of esters and terpenes, mainly *l*-limonene and cineole (Fig. 27).[2,3] The optically isomeric form of carvone, *d*-carvone is found in oil of caraway and oil of dill. Carvone when present in a plant appears to co-occur with limonene.

KEY TERMS

optical isomers

REFERENCES

1. Madison Avenue Aromatherapy, http://www.ashburys.com/zz_madison.html
2. "Chemical Perfume Materials," http://www.sh.com/ie/chemical/chem006.htm
3. "The Essential Oil for November. Peppermint-midwest," http:/atlanticinstitute.com/DrP/ eo.htm

RELATED WEB SITES

"Essential Notes: Peppermint," Madison Avenue Aromatherapy Center, http://www.madison-avenue.com/aroma/Oils/Peppermint.htm

"Essential Oils in Toothpaste," Steven Pearce, Britannia Natural Products, *Manufacturing Chemist*, July 1988, http://www.cotpubco.demon.co.uk/cosweb/teeth.html

Identification of the Volatile and Semi-Volatile Organics in Chewing Gums by Direct Thermal Desorption, John J. Manura, Scientific Instrument Services, Inc., Short Path Thermal Desorption—Application Note No. 12—October 1992, http://www.sisweb.com/referenc/ applnote/ap12.htm

WHAT IS THE DIFFERENCE BETWEEN A SUNSCREEN AND A SUNBLOCK?

The invisible ultraviolet rays of the sun can cause immediate and long-term skin damage in the form of sunburn, rashes, premature wrinkling, and skin cancer. To avoid overexposure, we are encouraged to apply sunscreens and sunblocks to protect the health of our skin. Chemistry clearly distinguishes between these two formulations, and the chemical structure of these products dictates how well these materials perform.

THE CHEMICAL ESSENCE

Sunblocks are opaque substances such as zinc oxide, titanium dioxide, and iron oxide that protect by forming a shield on the skin that reflects and scatters incident radiation. In essence, sunblocks provide *physical* protection against sun exposure, including both visible and ultraviolet light. Sunscreens *chemically* absorb ultraviolet light in the top layer of the epidermis, protecting the underlying layers.

Visible light ranges in wavelength from 400 to 700 nm. The spectrum of ultraviolet light from the sun ranges in wavelength from 200 to 400 nm and is divided into three broad classifications.[1] UVA rays have the longest wavelength (320 to 400 nm), are fairly constant year-round, and penetrate deeper into the layers of skin. Shorter UVB rays (290 to 320 nm) are more intense during the summer months than the longer wavelength UVA radiation. UVB radiation is also stronger at higher altitudes and in areas closer to the equator. UVC radiation, even shorter in wavelength from 200 to 290 nm, is absorbed by the stratospheric ozone layer and does not reach Earth's surface. (As ozone is depleted in the stratosphere, however, the range of wavelengths of UV light that reaches Earth's surface will become a greater concern for skin exposure.) The particular chemical structure of a sunscreen determines the wavelengths of ultraviolet light preferentially absorbed by a sunscreen.

THE CHEMICAL SPECIFICS

The sunblocks zinc oxide, titanium dioxide, and iron oxide are inorganic chemicals that are not absorbed into the skin. These substances consist of opaque particles that reflect both visible and ultraviolet light. In addition, zinc oxide blocks virtually the entire UVA and UVB spectrum[1] and thus offers overall protection. The particulate nature of these sunblocks enhances their effectiveness at reflecting sunlight. The smaller the particle size, the greater the surface area available for reflection, and the more effective the sun protection offered by the formulation.[2]

Sunscreens are transparent organic substances that penetrate into the skin and absorb ultraviolet radiation. Common classes of sunscreens include benzophenones, PABA derivatives, cinnamates, salicylates, and dibenzoyl-methanes.[3] The primary protective range of benzophenones is in the UVA region and include oxybenzone (Fig. 32), 270 to 350 nm; dioxybenzone (Fig. 33), 206 to 380 nm; and sulisobenzone (Fig. 34), 250 to 380 nm. PABA and PABA esters are primarily protective in the UVB range (290 to 320 nm), including PABA (Fig. 35), 260 to 313 nm; Padimate O (Fig. 36, also known as octyldimethyl PABA), 290 to 315 nm; Padimate A, 290 to 315 nm; and glycerol aminobenzoate (Fig. 37), 260 to 315 nm. Cinnamates are derivatives of cinnamon and are primarily protective in the UVB range (290 to 320 nm). Examples include octyl methoxycinnamate (Fig. 38), 280 to 310 nm, and cinoxate (Fig. 39), 270 to 328 nm. Salicylates, protecting in the UVB range (290 to 320 nm), include homosalicylate (Fig. 40), 290 to 315 nm; ethylhexyl salicylate (Fig. 41), 260 to 310 nm; and triethanolamine salicylate (Fig. 42), 269 to 320 nm. Dibenzoylmethanes serve best as protectors for the UVA range (320 to 400 nm), offering no protection from UVB. These substances include 4-tert-butyl-4'-methoxydibenzoyl-methane (Fig. 43), 310 to 400 nm, and 4-Isopropyldibenzoylmethane (Fig. 44), 310 to 400 nm.

What do all of these molecules have in common that enhances their ability to absorb ultraviolet light? One common structural element is the presence of an aromatic or benzene ring structure. The aromatic ring is the *chromophore* of sunscreens, that is, the functional group of atoms capable of absorbing ultraviolet light. Benzene, C_6H_6, absorbs in the ultraviolet at 280 nm. The presence of certain substituents on the benzene ring alters the electron distribution in the ring and shifts the absorption wavelength. In addition, the *conjugation of double bonds* (i.e., the presence of alternating double and single bonds as in $C=C-C=C$ or $C=C-C=O$)

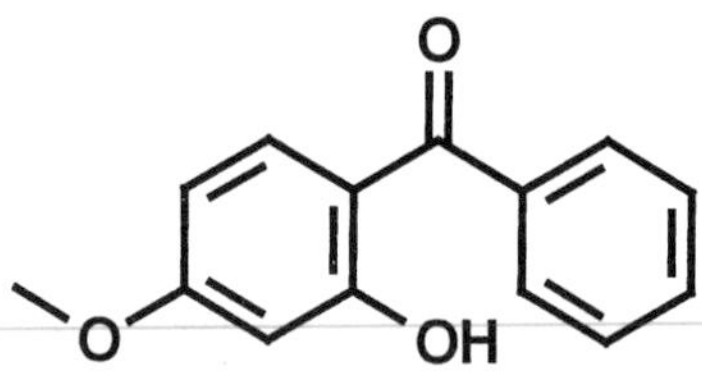

Figure 32   The chemical structure of oxybenzone.

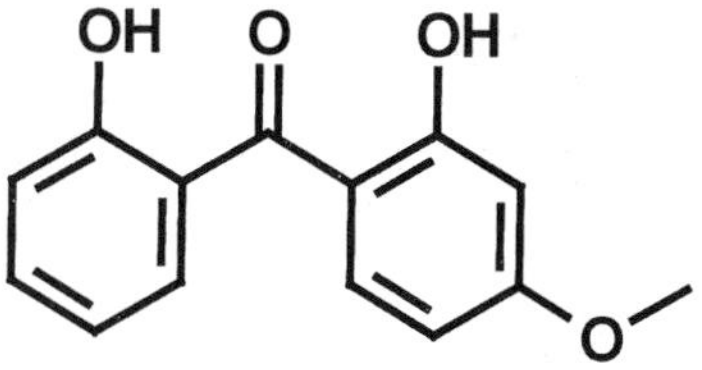

Figure 33   The chemical structure of dioxybenzone.

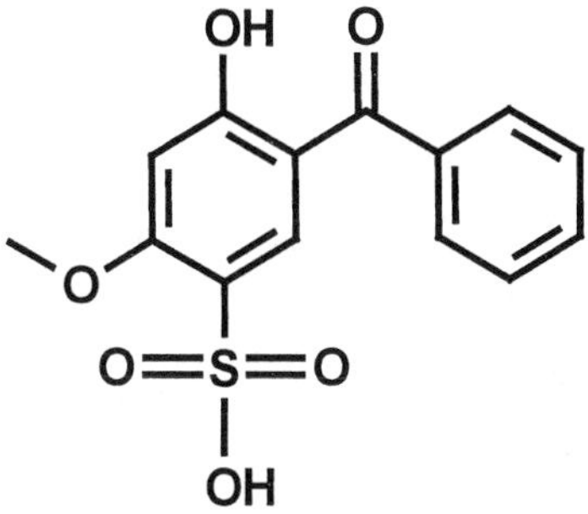

Figure 34   The chemical structure of sulisobenzone.

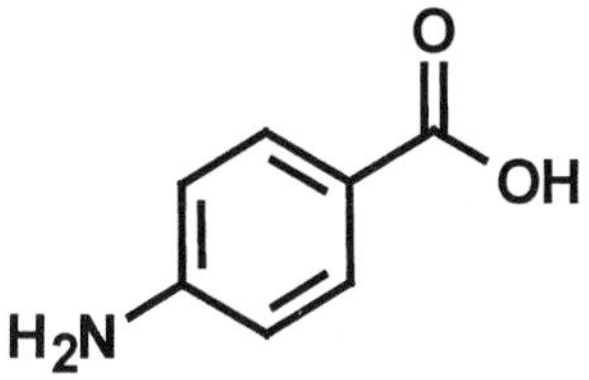

Figure 35   The chemical structure of PABA.

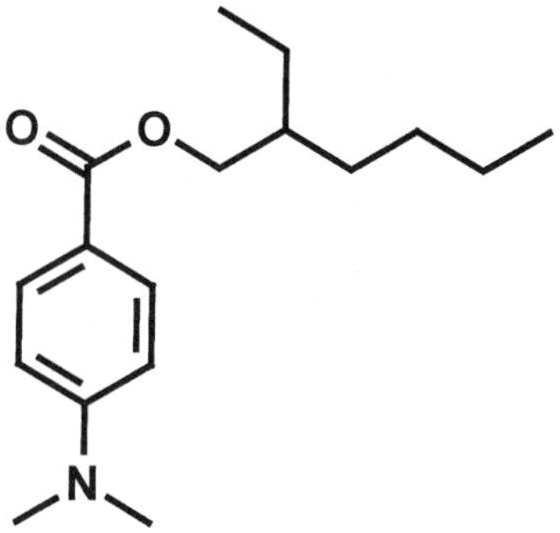

Figure 36   The chemical structure of Padimate O (or octyldimethyl PABA).

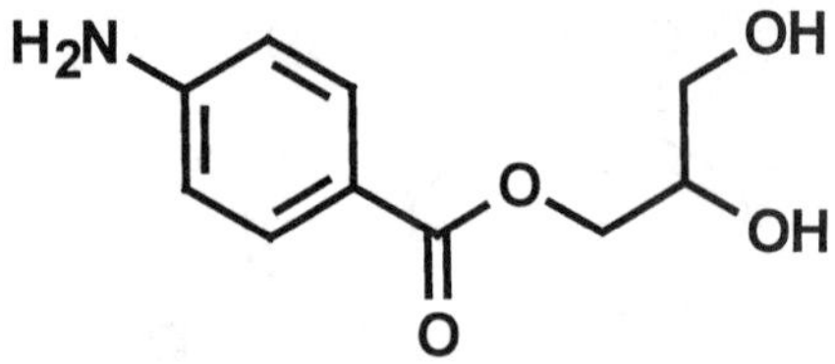

Figure 37   The chemical structure of glycerol aminobenzoate.

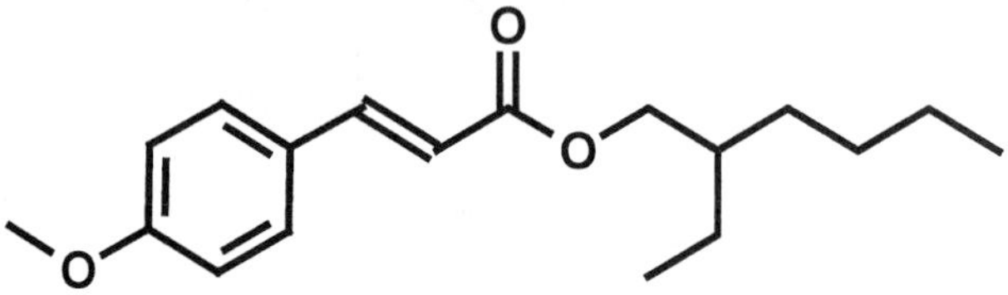

Figure 38   The chemical structure of octyl methoxycinnamate.

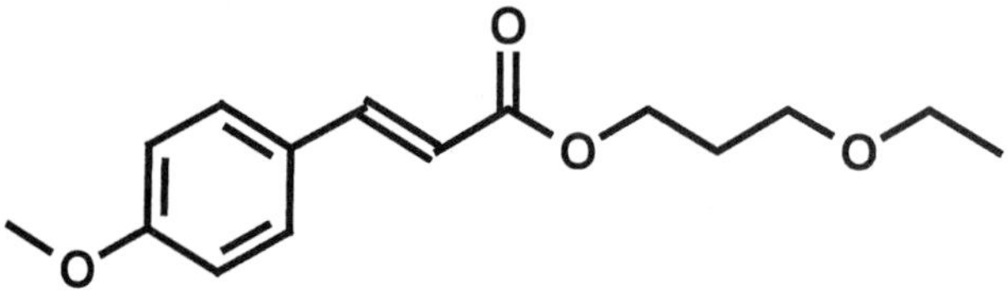

Figure 39   The chemical structure of cinoxate.

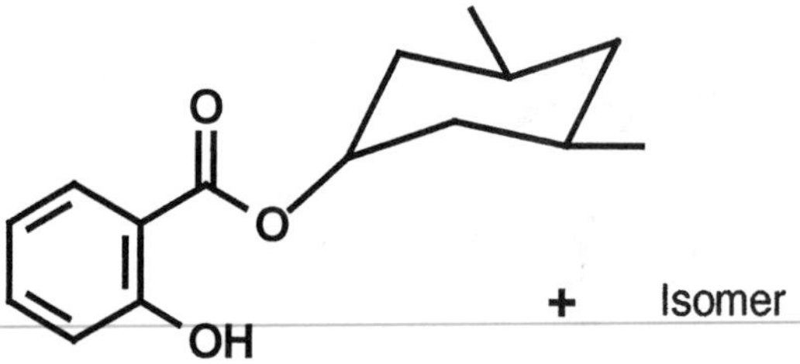

Figure 11.40   The chemical structure of homosalicylate.

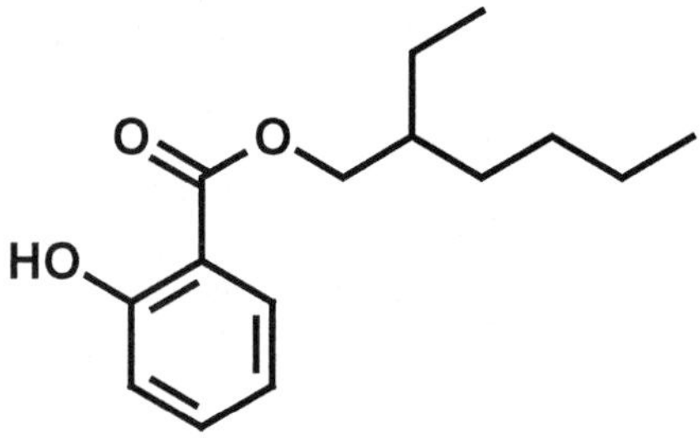

Figure 41 The chemical structure of ethylhexyl salicylate.

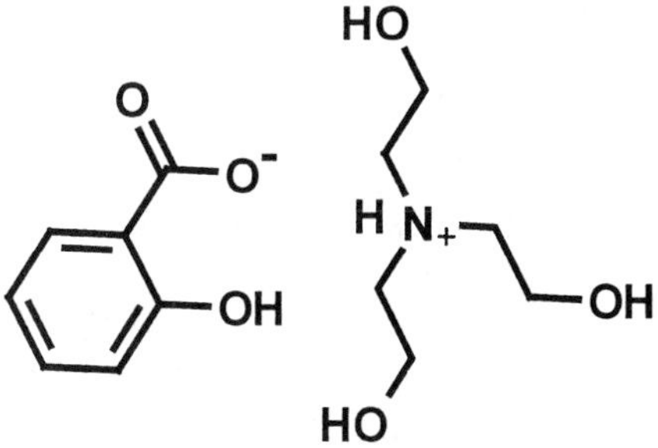

Figure 42 The chemical structure of triethanolamine salicylate.

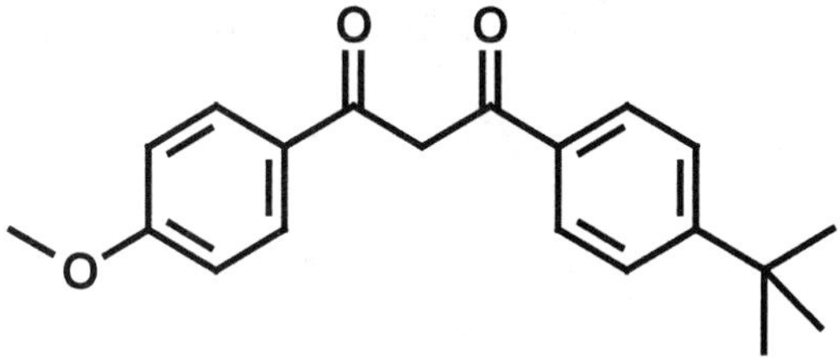

Figure 43 The chemical structure of 4-tert-butyl-4′-methoxydibenzoylmethane.

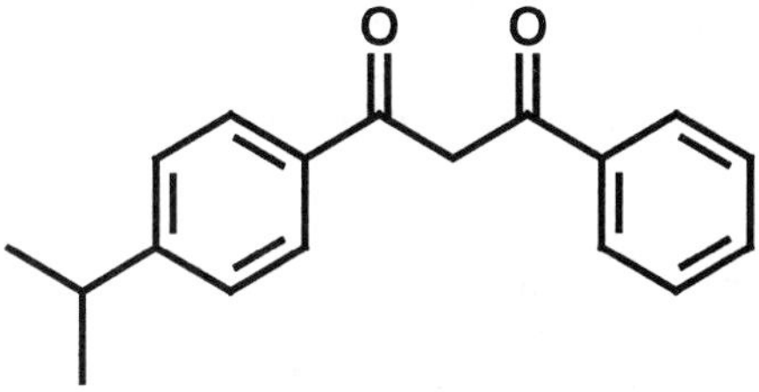

Figure 44 The chemical structure of 4-isopropyldibenzoylmethane.

can lead to tremendous shifts in absorption wavelength. In the class of benzophenones, the presence of hydroxyl ($-OH$) groups on the aromatic rings and the additional conjugation shift the absorption to longer wavelengths, particularly in the UVB range. Amino substituents ($-NH_2$) and carboxyl groups ($-COOH$) also increase the wavelength of absorption. Sunscreens thus have a chemical structure that matches their function—the absorption of ultraviolet light.

KEY TERMS

ultraviolet; chromophore; conjugation

REFERENCES

1. "What is Ultraviolet Light?" http://www.am.qub.ac.uk/users/j.pelan/uv/node2.html
2. "TiO₂sperse: Micronized Titanium Dioxide," http://www.collabo.com/tios.htm
3. "Sunscreens," http://www.geocities.com/HotSprings/4809/sunscr.htm#ACTIVE INGREDIENTS IN SUNSCREENS

WHAT MAKES A NO-TEARS SHAMPOO?

Shampoo manufacturers provide a wealth of products to satisfy a range of consumer preferences. What kind of chemical ingredients are included to ensure a mild shampoo and guarantee a "no-tears" formula?

THE CHEMICAL ESSENCE

As the central function of a shampoo is to cleanse the hair, the primary ingredient of a shampoo is a *detergent* (also known as a *surfactant*). Many shampoos, particularly those targeted for babies and children, claim to cause no eye irritation or sting. A "no-tears" formulation achieves this claim by carefully adjusting the nature of the surfactants. In particular, the identity and concentration of surfactants with an ionic or "charged" portion are controlled to minimize both eye and skin irritation.

THE CHEMICAL SPECIFICS

A surfactant is a molecule that is characterized as *amphiphilic*, that is, containing both a discrete *hydrophilic* (water-soluble) or polar portion and a well-defined *hydrophobic* (oil-soluble) or nonpolar fragment. The hydrophilic portion of the molecule is called the surfactant *headgroup;* the hydrophobic portion of a surfactant is described as the surfactant *tail.* The tail group is generally a linear long-chain hydrocarbon residue, such as the linear dodecyl group, $-C_{12}H_{25}$. Other hydrophobic groups are possible, including substituted benzene and other aromatic rings. The polar headgroup is generally classified as one of four types, depending on the charge on the hydrophilic group; anionic, cationic, nonionic, and amphoteric or zwitterionic headgroups are possible.

Anionic surfactants carry a negatively charged headgroup and are extremely useful in shampoo because of their excellent cleansing, foaming, and water solubility properties. Anionic surfactants also rinse easily from the hair. These surfactants are also relatively inexpensive and easy to synthesize. For these reasons, anionic surfactants are the most common surfactants in personal care products. Nevertheless, anionic surfactants are known to be harsh and irritating to both eyes and scalp. Alkyl sulfates are the most frequently used anionic surfactants, including sodium (Fig. 45), ammonium (Fig. 46), and triethanolammonium (TEA) (Fig. 47) lauryl sulfates.

Cationic surfactants contain a positively charged headgroup and are typically used as conditioners to improve hair manageability and reduce static. Cationic surfactants are especially irritating to eyes when used in high concentrations but are safe and useful in low amounts. Quaternium-15 (chloroallyl methanamine chloride, Fig. 48) is a cationic surfactant included in shampoo formulations for its conditioning ability.

Nonionic (i.e., neutral) surfactants are not primarily used for their cleansing properties but to improve the solubility, foaming action, and conditioning action of a shampoo formulation. One nonionic detergent that is especially effective at reducing eye irritation is TWEEN 20 Polysorbate 20 (Fig. 49).[1]

Amphoteric or zwitterionic surfactants (containing both positive and negative charges) are used for their low foaming and slight irritation characteristics. Gentle, "no-tears" shampoos contain significant amounts of am-

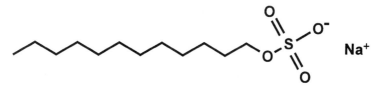

Figure 45   The chemical structure of the anionic surfactant sodium lauryl sulfate.

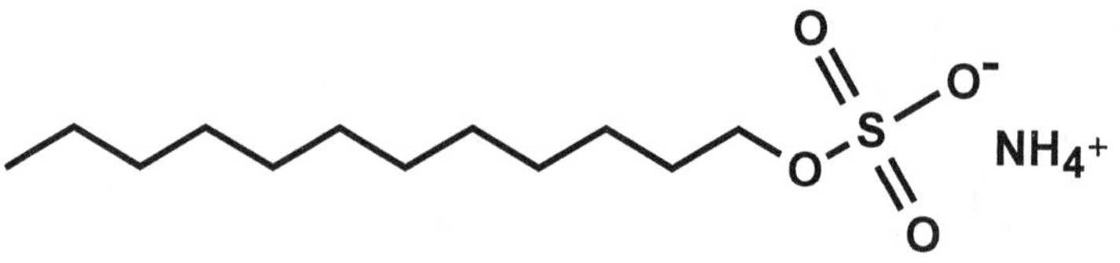
O
O⁻
S
O
O
NH4+

Figure 46 The chemical structure of the anionic surfactant ammonium lauryl sulfate.

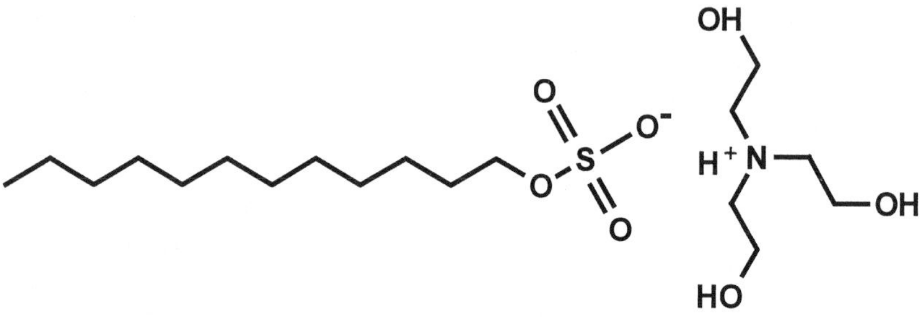
OH
O
O⁻
S
O
O
H+ N
OH
HO

Figure 47 The chemical structure of the anionic surfactant triethanolammonium (TEA) lauryl sulfate.

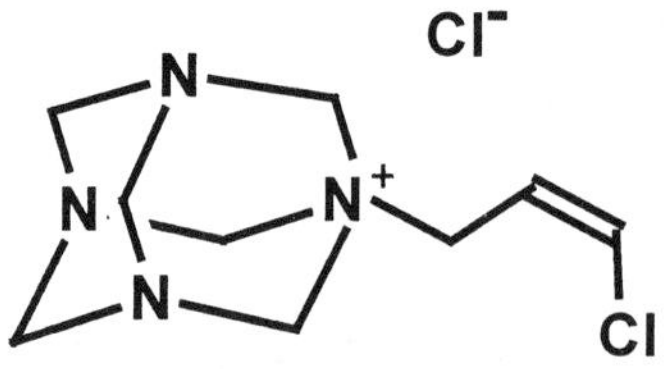
Cl⁻
N
N
N+
N
Cl

Figure 48 The cationic surfactant quaternium-15 (chloroallyl methanamine chloride).

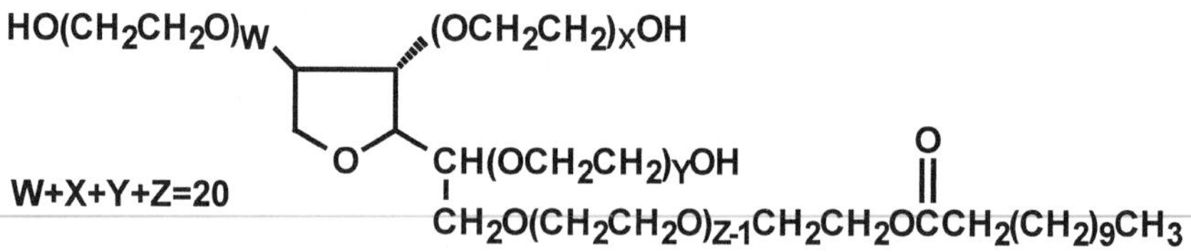
HO(CH2CH2O)W
(OCH2CH2)XOH
O
W+X+Y+Z=20
CH(OCH2CH2)YOH
O
CH2O(CH2CH2O)Z-1CH2CH2OCCH2(CH2)9CH3

Figure 49 The anionic surfactant TWEEN 20 Polysorbate 20.

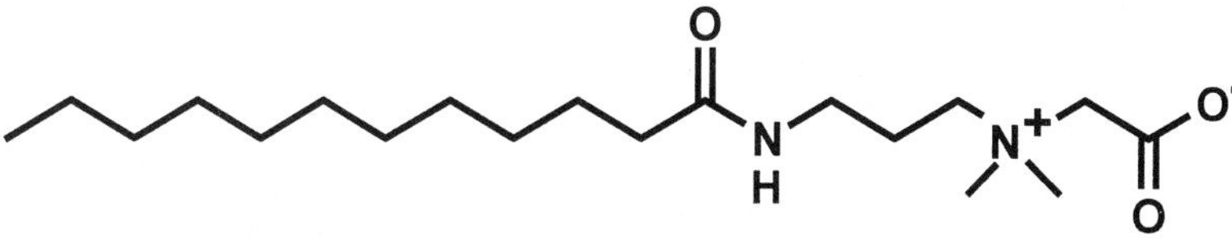

Figure 50 The amphoteric surfactant lauramidopropyl betaine.

photeric surfactants. An example of an amphoteric surfactant found in shampoo is lauramidopropyl betaine (Fig. 50).

The net charge on the headgroup of a surfactant is one parameter that affects eye irritation. Experimental studies suggest that a relationship exists between eye irritation (I) and surfactant concentration (C) of the form $I = kC^2$, where k is a constant for a given surfactant or combination of surfactants and C is the concentration of individual surfactant molecules.[2]

KEY TERMS

surfactant

REFERENCES

1. "Shampoos," ICI Surfactants, http://www.surfactants.com/Shampoos.html
2. "The Practical Importance of Surfactant Monomer Concentrations," Eric Lomax, http://alphaline.com/11/SPECCHEM.htm

RELATED WEB SITES

"Chemical Fact Sheet—Anionic Surfactants," Orica Limited, http://www.orica.com.au/business/cor/wcor00013.nsf/Webnav2/Chemical+Fact+Sheets?OpenDocument
"Chemical Fact Sheet—Cationic Surfactants," Orica Limited, http://www.orica.com.au/business/cor/wcor00013.nsf/Webnav2/Chemical+Fact+Sheets?OpenDocument
"The Structure of Your Hair," http://www.geocities.com/HotSprings/4266/chem.html
"What Is in a Shampoo," Chemistry Factoids, California State University, Fresno, http://129.8.104.30:8080/projects97/115.html

WHAT CAUSES THE FIZZ WHEN AN ANTACID TABLET IS ADDED TO WATER?

The familiar fizzing action that occurs as an antacid tablet dissolves in water is the result of a chemical reaction involving the ingredients in the tablet.

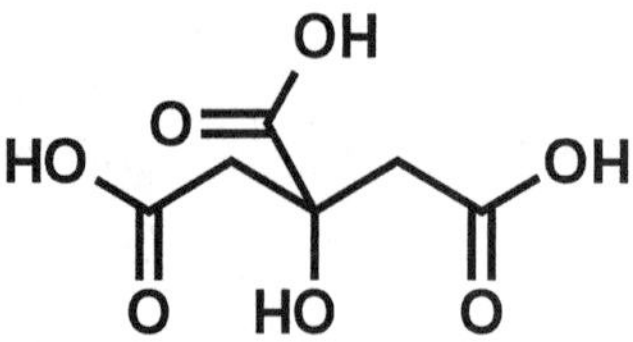

Figure 51   The chemical structure of citric acid.

THE CHEMICAL ESSENCE

The effervescence of an antacid tablet is water is key to the effectiveness of the antacid. Two familiar antacids, Alka Seltzer and Bromo Seltzer, contain sodium bicarbonate or baking soda $-NaHCO_3$. The bicarbonate ion reacts with citric acid, another ingredient of the product, to produce carbonic acid, H_2CO_3. In solution, carbonic acid decomposes to yield water and gaseous carbon dioxide (CO_2), the same gas in carbonated beverages and champagne. The fizz generated when the tablet dissolves is simply the CO_2 bubbles coming out of solution.

THE CHEMICAL SPECIFICS

Each Alka-Seltzer tablet contains 1916 mg of sodium bicarbonate, 1000 mg of citric acid (Fig. 51), and 325 mg of aspirin or acetylsalicylic acid (Fig. 52).[1] Bromo Seltzer also contains sodium bicarbonate and citric acid as well as acetaminophen (Fig. 53).

Several equilibria describe the action of bicarbonate-based antacid tablets. First of all, sodium bicarbonate dissolves completely in aqueous solution to generate sodium ions and bicarbonate ions:

$$NaHCO_3 \text{ (s)} \rightarrow Na^+ \text{ (aq)} + HCO_3^- \text{ (aq)}$$

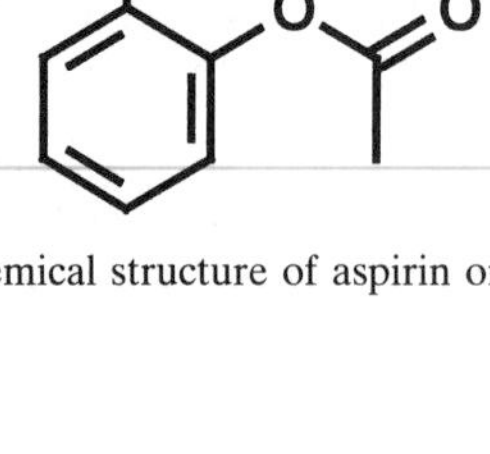

Figure 52   The chemical structure of aspirin or acetylsalicylic acid.

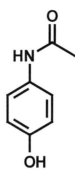

Figure 53 The chemical structure of acetaminophen.

In water the bicarbonate ion is in equilibrium with carbonic acid:

$$HCO_3^- \text{ (aq)} + H_2O \text{ (l)} \rightarrow H_2CO_3 \text{ (aq)} + OH^- \text{ (aq)}$$

$$K_{eq} = K_a \text{ (}HCO_3^-\text{)} = 5.6 \times 10^{-11}$$

The small value of the equilibrium constant indicates that the formation of carbonic acid is not very extensive in neutral water. However, the formation of carbonic acid is quite favored in acidic solution (arising from the citric acid also contained in the product):

$$HCO_3^- \text{ (aq)} + H_3O^+ \text{ (l)} \rightarrow H_2CO_3 \text{ (aq)} + H_2O \text{ (l)}$$

$$K_{eq} = 1/K_a \text{ (}H_2CO_3\text{)} = 1/(4.3 \times 10^{-7}) = 2.3 \times 10^6$$

The carbonic acid produced via the reaction with stomach acid decomposes to yield dissolved gaseous CO_2:

$$H_2CO_3 \text{ (aq)} \rightarrow H_2O \text{ (l)} + CO_2 \text{ (g)}$$

Production of H_2CO_3 by the action of bicarbonate ions on the acid in the stomach drives the preceding equilibrium to the right to generate CO_2 bubbles.

KEY TERMS

equilibrium; equilibrium constant

REFERENCE

1. "Alka Seltzer Original: Back of Package Information," Bayer Corporation, http://www.alka-seltzer.com/alka_prods/alka_cluster_fr.htm

RELATED WEB SITES

"Alka-Seltzer Presents Cool Science Experiments," Bayer Corporation, http://www.alka-seltzer.com/experiments/experiments.htm

WHAT IS THE DIFFERENCE BETWEEN HARD AND SOFT CONTACT LENSES?

The number of options for contact lens wearers has been transformed by innovative advances in the chemistry of the polymeric materials used to create these vision products.

THE CHEMICAL ESSENCE

Most everyone is familiar with the contrasting resilient, pliable nature of soft contact lenses and the brittle character of hard lenses. Both lenses are constructed from polymers, but the differing chemical composition of each polymer leads to considerably different physical properties.

Hard contact lenses are composed of a polymer that repels water because the constituent repeating units (the *monomers* that link together to form the polymer) are nonpolar, *hydrophobic* segments. The first hard contact lens was constructed in 1948 from the monomer known as methyl methacrylate (MMA), yielding the polymer poly(methyl methacrylate) or PMMA. This material offers durability, optical transparency, and acceptable wettability for optimal comfort. Today the rigid lens material of hard contact lenses is often constructed by combining MMA with one or more additional hydrophobic monomers to provide better gas permeability.

The first soft contact lenses were also constructed with a polymeric material containing a single monomeric unit. The added pliability of the soft lens was derived from the more hydrophilic nature of the monomer, enhancing the ability of the polymer to absorb water and provide greater comfort to the lens wearer. This monomer is a derivative of MMA known as hydroxyethyl methacrylate (HEMA). A number of hydrophilic monomers are used in soft lenses today; these materials are referred to as *hydrogels* because of their ability to absorb significant amounts of water yet remain insoluble.

The soft extended wear lenses popular today are composed of polymers with more than one type of repeating unit (i.e., *copolymers*). For extended wear a lens with greater oxygen permeability is needed, for the cornea relies on direct oxygen transmission from the atmosphere as a consequence of the lack of blood vessels within the corneal framework. Scientists have found that the higher the water content of a hydrogel polymer, the more extensive the oxygen permeability of the lens formulated from that polymer. A 70% water content is desirable for extended periods of lens wear, expressed as a percentage of the total weight of the polymer. To enhance the water content of these lenses, two basic formulations are employed. In one case a hydrophilic monomer such as HEMA is combined with a highly hydrophilic charged (ionic) monomer. As a consequence of the strong attraction interactions between water and the charged functionality, the water content of the lens is greatly increased, increasing the pliability and comfort level of the lens. Alternatively, the lens can also be designed by combining two highly hydrophilic nonionic monomers.

THE CHEMICAL SPECIFICS

Cross-linked polymeric materials with optical transparency and biocompatibility are used to construct hard contact lenses. The monomers commonly used in hard contact lenses possess a high degree of hydrophobicity due to their inability to form hydrogen bonds with water. The ester methyl methacrylate (MMA) (Fig. 54), $CH_2C(CH_3)COOCH_3$, was the first monomeric unit used in 1948.

Lenses with a greater degree of gas permeability were designed in the mid 1970s using siloxane-based monomers. For example, a copolymer of methyl methacrylate and the monomer known as methacryloxypropyl tris(trimethylsiloxy silane) or TRIS (Fig. 55) was formulated in 1975 to provide a number of desirable features such as oxygen permeability, wettability, and scratch resistance.[1] In the 1980s, approaches using a number of fluorine-based monomers were successful. A polymer of MMA, TRIS, and hexafluoroisopropylmethacrylate (HFIM) (Fig. 56) is one such formulation for hard lenses.

The hydrogel poly(hydroxyethyl methacrylate) composed of cross-linked monomers of 2-hydroxyethyl methacrylate (HEMA) (Fig. 57) was the first

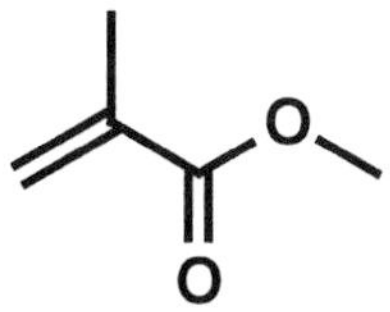

Figure 54   The chemical structure of the ester methyl methacrylate (MMA).

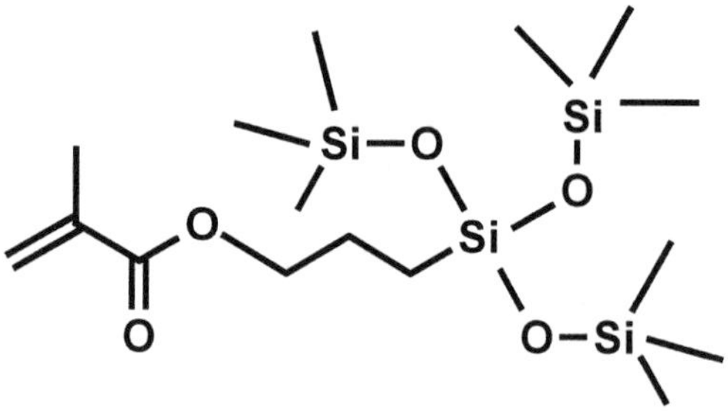

Figure 55 The crystal structure of the monomer methacryloxypropyl tris(trimethylsiloxy silane) or TRIS.

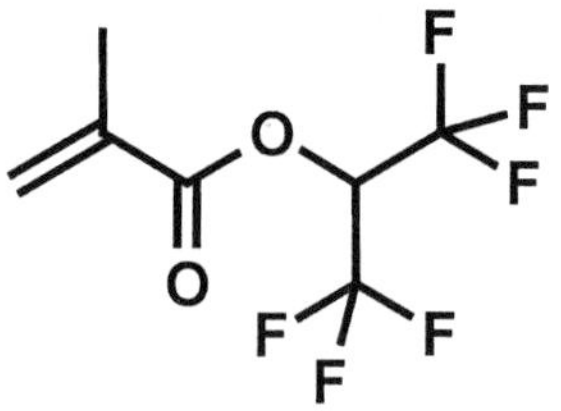

Figure 56 The chemical structure of the monomer hexafluoroisopropylmethacrylate (HFIM).

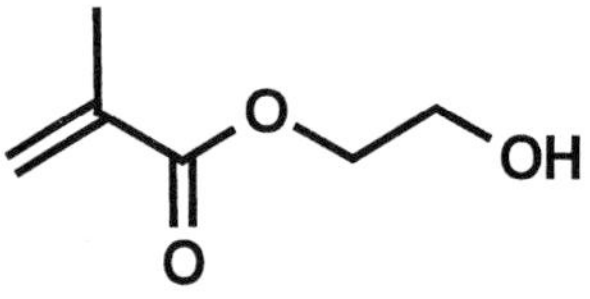

Figure 57 The chemical structure of the monomer 2-hydroxyethyl methacrylate (HEMA).

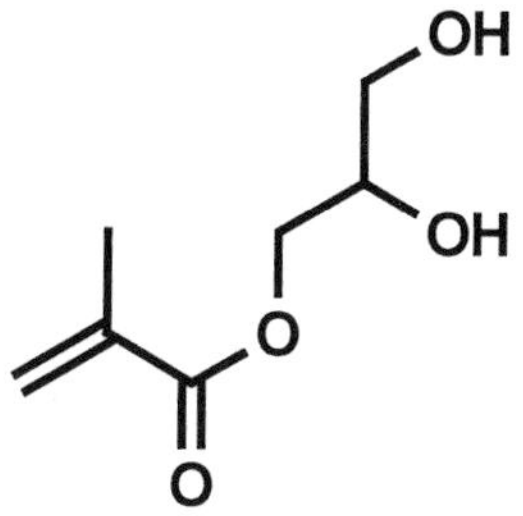

Figure 58 The chemical structure of the hydrophilic nonionic monomer 2,3-dihydroxy-propylmethacrylate.

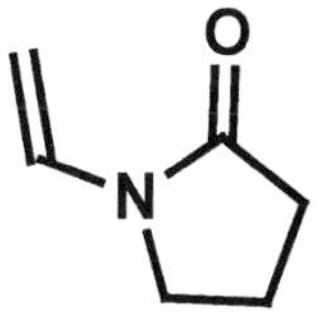

Figure 59 The chemical structure of the hydrophilic nonionic monomer N-vinyl-2-pyrrolidinone (NVP).

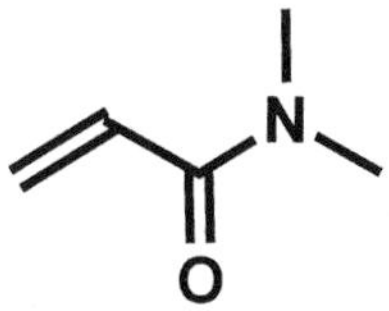

Figure 60 The chemical structure of the hydrophilic nonionic monomer N,N-dimethyl-acrylamide (DMA).

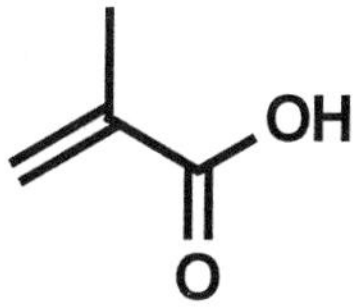

Figure 61 The chemical structure of the monomer methacrylic acid (MAA).

soft lens material. (*Cross-linking* consists of bonding between the main polymer chains to add strength to the material.) The presence of the hydroxy functional group (i.e., —OH) permits hydrogen bond formation with water and leads to the capacity to absorb water (typically 38% by weight).[2] Highly hydrophilic nonionic monomers used in hydrogel lenses include glycerol methacrylate (GM) or 2,3-dihydroxypropylmethacrylate (Fig. 58), N-vinyl-2-pyrrolidinone (NVP) (Fig. 59), and N,N-dimethylacrylamide (DMA) (Fig. 60). The Accuvue lens currently manufactured by Johnson & Johnson is an example of a hydrogel polymer with an ionizable monomer (hence enhanced water absorption). Cross-linking HEMA (Fig. 57) and methacrylic acid (MAA) (Fig. 61) leads to a water content of 58%.

KEY TERMS

hydrophobic; hydrophilic; monomer; polymer; copolymer; hydrogel; cross-linking

REFERENCES

1. "Contact Lens Material," Dr. Jay F. Kunzler and Dr. Joseph A. McGee, Department of Polymer Chemistry, Bausch & Lomb, *Chemistry & Industry,* http://ci.mond.org/9516/951608.html
2. "Contact Lens Products," CIBA Vision, http://www.cibavision.com/text/onsight/ourproducts/A02.01.html

RELATED WEB SITES

"Contact Lenses Identified by Type of Material," CIBA Vision, http://www.cibavision.com/text/forsight/allabout/C04.03.05.html

Index

A

Absorber material, 160
Absorption, and hydrogen bonding, 177
Absorption spectrum, versus emission
 spectrum, 183
Acetaminophen, 210
 chemical structure, 211
Acetic acid, 98
Acetylsalicylic acid, chemical structure, 210
Acid/base content, 73; *see also* pH
 indicators of, 83–84
Acid rain, 100
Acrylics, 97
Adsorption, 124
 water, to silica microspheres, 168
Albumen, 48
Alcohols; *see* also Ethanol
 polyfunctional, 148
 polyhydric, in cold creams, 188
Alkali metals, thermal ionization, 128
Alkalinity, 113; *see also* pH
Alkyl sulfates, 207
Alloys, 99
 fusible, 85–86
 liquid metal, 30, 31
 oxidation of, 140
Alpha hydroxy acids (AHAs)
 in antiwrinkle creams, 189
 chemical structure, 189
Alpha hydroxycarboxylic acid, 111
Alumina, 55, 56
Aluminosilicate glass, 91
Aluminum, and hydrangea color, 64
Aluminum oxide, 139

Amidotrizoates, 8
α-Amino groups, 162
7-Amino-4-methylcoumarin, 183
 chemical structure and absorption/
 emission spectra, 182
Aminophenol, 196
 chemical structure, 195
Amino substituents, 206
Ammonia, 95, 96, 97
Ammonium lauryl sulfate, 79, 207
 chemical structure, 80, 208
Ammonium perchlorate, 55
Anesthetics, concentration in brain, factors,
 13–14
Anionic surfactants, 207
Anode, 140
Anthocyanins, 64
 chemical structure, 65
 in basic solution, 66
Anti-caking agents, 36
Antistatic compounds, 108
Aramid fibers, 172
Aromatic ring structure, 202
Arrhenius, Savante, 73
Arrhenius law, 25
Artificial hip, design of, 18
Ascorbic acid, 189
 chemical structure, 190
Aspirin, *see* Acetylsalicylic acid
Aurora borealis/australis, 60

B

Baker's yeast, 37
Baking soda, *see* Sodium bicarbonate

Barium sulfate, 6
Basic lead sulfate, 142
Benzene, 202
Benzene ring structure, 202
Benzophenones, 206
Beta hydroxy acids (BHAs), 189, 191
    chemical structure, 190
Biomaterials, 18
Biphenyl, 180
    chemical structure, 182
Bismuth, melting point, 85
Bisphenol A
    chemical structure, 114
    synthesis, 115–116
9,10-Bis(phenylethynyl)anthracene, 26
    chemical structure, 27
Bleach, household, 95
Bone, composition of, 19
Borax, 112
Borosilicate glass, 127
Brass, surface corrosion, 99
Bromine, in halogen lights, 89
Bubbles
    in colloidal dispersion, 48
    spherical shape of, 46
Bunsen, Robert, 138
Butane, 71
1,4-Butanediol, *see* Butylene glycol
2-Butene-1-thiol, chemical structure, 69
Butylene glycol, 193
    chemical structure, 195
tert-Butyl mercaptan, chemical structure, 72

C

Cadmium, melting point, 85
Calcium carbonate, 62, 98, 110, 167
Calcium chloride, deliquescent nature
    of, 124
Calcium hydroxyapatite, 19
Calcium hypochlorite, 29
Calcium oxide (lime)
    crystallization, 136
    as desiccant, 123
    incandescence of, 135
Calcium silicate, 36
Calcium sulfate, as desiccant, 123, 124
Capillary condensation, 124
Capric acid, 110
Caprylic acid, 110
Capsaicin, chemical structure, 34

Capsaicinoids, 34
Carbonate white lead, 142
Carbonation, process of, 44
Carbon dioxide
    in cloud seeding, 58
    in effervescence of antacids, 210
    as fire extinguisher, 102, 103
    introduction into water, 44
    solubility in water, 45
Carbonic acid, 45, 83, 210
    ionization of, 84
Carboxyl groups, 206
l-Carvone, 200
    chemical structure, 199
Cation exchange, in water softeners,
    110–111
Cationic surfactants, 207
CAT scans, *see* Computerized axial
    tomography
Cellulose, 2
    repeating monomeric units, 171
    structure, and hydrogen bonding, 178
Cetyl alcohol, chemical structure, 193
CFC-113, 119
    chemical structure, 120
Chain-growth polymerization reaction, 79
Chain-transfer agent, 79
Chelation, 43
Chelators, 111
Chemiluminescence, 23
Chemoreception, 197
Chloramines, 95
Chlorine, 28
Chloroallyl methanamine chloride, *see*
    Quaternium-15
Chlorofluorocarbons, 119–120
Chloroform, 10
Chromatophores, 63
Chrome pigments, 142
Chromic acid, 185
Chromophores, 202
Cineole, 197
    chemical structure, 199
Cinnamates, 202
Cinoxate, 202
    chemical structure, 204
Citric acid, 43, 189, 192
    chemical structure, 190, 210
Citrus crops, 67
Clausius–Clapeyron equation, 38, 39

Cloud condensation nuclei, 57
Cloud seeding, 57–59
Coacervation, 76–77
Coalescence, 57
Coefficient of thermal expansion, 126
Cold stabilization, of wine, 51
Collagen, 19
Colligative property, 132
Colloidal dispersion, 48
Combustion, 101–103
　components of, 101
　metal, 103–104
　zirconium, 139
Complex formation, in cyanide
　　poisoning, 21
Computerized axial tomography, 5, 7
Conalbumin, 48
Condensation, in raindrop formation, 57
Condensation reaction, polymers, 2
　in Kevlar synthesis, 173–174
Conjugate base, 84
Conjugation of double bonds, 202, 206
Contrast media, 6, 7
Coordination complex, 22
Coordination number, 136
Copolymers, 96, 212
Copper
　oxidation products, 100
　surface corrosion, 99
Copper resinate, 143
Cormack, Allen, 5
Correction fluids, 81–82
Cotton
　in currency fabric, 151
　dye interaction, and hydrogen
　　bonding, 185
　fibers, hydrogen bonding of, 177
Coumarin, 180
　chemical structure, 181
Crane & Company, 152
Cream of tartar, 48, 51
Cristobalite, 167
Cross-linking, polymer, 216
Crystals/crystallization, 31, 50
　calcium oxide, 136
　diamond, 165
　glass, 90
　gray tin, 146
　ice, 58
　mica, 121

in thermoplastics, 115
　white tin, 145
　wurtzite structure, 59
Currency
　cotton/linen fabric of, 151–152
　optically variable ink in, 158
　security thread in, 155
Cuttlefish (*Sepia officinalis*), 63
Cyanide poisoning, 20
　antidote for, 21–22
　mechanism, 21
Cyanoacrylates, 130
Cytochrome oxidase, 21

D

Dacron
　linked monomers, 179
　repeating unit, 178
Decomposition reactions, 38
Deliquescence, of calcium chloride,
　124
Density, 93
Desflurane, 12
Desiccants, 123
Diamines, 173
Diatrizoate, 8
Dibenzoylmethanes, 202
Dicarboxylic acid, 173
Dichloramine, 95
Dicobalt edetate, 21
Dielectric material, 160
16,17-Dihexyloxyviolanthrone, 26
　chemical structure, 28
Dihydrocapsaicin, 34
　chemical structure, 35
2,3-Dihydroxypropylmethacrylate, chemical
　structure, 215
N,N-Dimethylacrylamide (DMA), 216
　chemical structure, 215
Dioxybenzone, chemical structure, 203
9,10-Diphenylanthracene, 25–26, 26
Disappearing ink, 82–83
Dispersion, 147
Dispersion forces, 108
Disproportionation, 106
Di-(4-tert-butylcyclohexyl)peroxydicarbonate,
　chemical structure, 80
DMA, *see N,N*-Dimethylacrylamide
Dodecanethiol, chemical structure, 79
Double bonds, conjugation, 202, 206

Drain cleaners, 112–113
Dust particles, 107

E

Edison, Thomas, 87, 88
EDTA, *see* Ethylenediaminetetraacetic acid
Electrical charge, 107
Electron microscopy, 167
Emission spectrum, versus absorption
    spectrum, 183
Emulsions, solid, 167
Endothermic reactions, 102, 187
Enflurane, 12
Enteric coatings, 3
Enthalpy, 102
    of fusion, 67
    of vaporization, 38–39
Entropy of vaporization, 38–39
ePTFE, *see* Expanded
    polytetrafluoroethylene
Equilibrium constant, in formation of
    carbonic acid, 211
Essential oils, 197
Ester linkages, polyester-forming, 156
Ethane, 71
Ethanethiol, chemical structure, 71
Ethanol, 51
    chemical structure, 195
    in cold creams, 187–188
    flash point of, 49
    freezing/boiling points, 126
    hydroxyl groups in, 193
    use in thermometers, 125–126
    vapor pressure of, 188
Ether (diethyl ether), 14
    chemical structure, 10
Ethyl chloride, 10
    chemical structure, 11
Ethylenediaminetetraacetic acid (EDTA),
    21–22
    chemical structure, 41
    formation constants, 42
    as sequestrant in food products, 41–43
16,17-(1,2-Ethylenedioxy)violanthrone, 26
    chemical structure, 28
Ethylhexyl salicylate, 202
    chemical structure, 205
Evaporation, 187
Exfoliators, 189
Exothermic processes, 67

Exothermic reaction, 104
Expanded polytetrafluoroethylene, 171

F

Fabrics
    bullet-resistant, 172–176
    cotton/linen, of currency, 151–152
    denim, pigmentation of, 184–185
    water-repellent/waterproof, 169–171
Face-centered cubic lattices, 136
Fahrenheit, Daniel Gabriel, 126
Fatty acids, in soap, 110
Fermentation, 38
Ferric hydroxide, 143
Fibers, natural, yellow tinge of, 180
Fire
    components of, 101
    magnesium, 103–104
Fire sprinkler heads, automatic, 84–85
Flashbulbs, photographic, 104, 138–139
Flash point, 49
Fluorescence
    in optical brighteners, 180
    in security fibers of currency, 156
Fluorescent dyes, 25
Fluorocarbons, in water-resistant
    fabrics, 171
Fluorosilicic acid, 90
    chemical structure, 90
Foams, 48
Fog
    artificial production of, 147–148
    versus smoke, 147
Formation constants, EDTA, 42
Free-radical polymerization reaction, 79
Freezing point depression constant, 132
Freezing process, 67
Fry, Art, 78
Furfurylthiol, 69
Fuwa, Kyozo, 90

G

Galen, 15
Galileo, 125
Gas permeability, of contact lenses, 213
Glass
    reaction with hydrofluoric acid, 90
    shatterproof, 114
    thermal expansion of, 126–127
Glassy metals, 30

Glycerin, 193
  chemical structure, 195
Glycerol aminobenzoate, 202
  chemical structure, 204
Glycerol methacrylate, 216
Glycolic acid, 111, 189, 192
  chemical structure, 16, 190
Gold films, 54
Groundwater, 98, 109–110

H

Hailstones, 58
Hale-Bopp, Comet, 128
Halogen lights, 89
Halothane, 10
  chemical structure, 11
Hard water, 109
Heat capacity, 68
HEMA, *see* 2-Hydroxyethyl methacrylate
Henry's law, 45
Heterochain polymers, 114
Heterogeneous reaction, 102–103
Hexadecanol, *see* Cetyl alcohol
Hexafluoroisopropylmethacrylate
      (HFIM), 213
  chemical structure, 214
1,6-Hexamethlenediamine, 173
Homosalicylate, 202
  chemical structure, 204
Hope diamond, 164
Hounsfield, Godfrey, 5
Hydrangeas, 63
Hydrazine, 95
Hydrocarbons, 71
  dispersion forces of, 108
  as fire component, 101
  halogenated, 119
Hydrofluoric acid, 90
Hydrogen bonding, 33, 47
  of cellulose, components of, 178–179
  of cotton fibers, 173
  of hydrophilic substances, 119
  in Kevlar synthesis, 173, 175, 176
  loss of, in water-resistant fabrics, 170
  of polyfunctional alcohols, 148
  in polymer–surface bonding, 130
Hydrogen peroxide, 104
  chemical structure, 105
Hydrogen sulfide, 69, 71, 142
Hydrolysis, 17

Hydrophobic force, 93
Hydrophobicity
  of hard contact lenses, 213
  of surfactant molecule, 207
Hydroxyaniline, *see* Aminophenol
2-Hydroxyethyl methacrylate (HEMA),
      212, 216
  chemical structure, 213
Hydroxyl groups
  in alcohols, 193
  in alpha hydroxy acids, 189
  in benzophenones, 206
  in cellulose-based fibers, 170
  in leucoindigo, 185
Hygroscopicity, 147–148

I

Ideal gas law, 39–40
Imidazole, 180
  chemical structure, 181
Incandescence, 135
Indigo dye, 184
Indium, melting point, 85
Infrared radiation, 54
Intermolecular forces, 177
Iodine, 6
  in halogen lights, 89
Ions/ionization, 84
  aluminate, 84
  atmospheric gases, 60–61
  carbonate, in formation of calcium
      carbonate, 98
  hydrogen, and pH scale, 73
  metal
      and disproportionation process, 106
      in soap scum formation, 110
  thermal, alkali metals, 128
Iothalamates, 8, 9
Iridescence, 166
Iron, standard reduction potential, 141
Iron dioxide, in sunblocks, 201
Iron oxide, 121, 140
Isoflurane, 12
  chemical structure, 13
Isooctyl acrylate, chemical structure, 79
4-Isopropyldibenzoylmethane, 202
  chemical structure, 205

K

Ketamine hydrochloride, 10
  chemical structure, 11

Ketones, 198
Kevlar, repeating unit and hydrogen
    bonding, 175

L

Lactic acid, 189
    chemical structure, 16, 190
Langmuir, Irving, 58
Lapis lazuli, 142
Lattices
    face-centered cubic, 136
    of diamonds, 165
    in zeolite structure, 124
Lauramidopropyl betaine, chemical
    structure, 209
Lauric acid, 110
Lauric acid diethanolamine, 196
    chemical structure, 195
Lava lamp, 92–94
Lead
    melting point, 85
    in pigments, 142
Leavening agents, 37
Leucoindigo, chemical structure, 185
Light
    absorption versus emission spectra, 183
    color diffraction phenomenon, 159
    and disproportionation process, 106
    interference, and iridescence, 166
    polarization of, 198
    refraction
        in paper, 153
        by solutes in ice, 132
    refraction/reflection, in pearlescent
        pigments, 121–122
    from sodium ionization, 128
    ultraviolet, classifications of, 201
    wavelengths in auroras, 61
Lightbulbs; *see also* Flashbulbs
    frosted, 90
    incandescent, 87–88, 135
        versus halogen bulbs, 89
Light sticks
    chemical reaction in, 25–26
    uses of, 24
Lime, calcinated, deliquescent nature
    of, 124
Limelights, 135–136
Limestone, thermal decomposition of, 136
Limonene, chemical structure, 108, 109

l-Limonene, chemical structure, 200
Linoleic acid, 110
Liquidmetal, 30
Liquid-to-gas phase transition, 12
Logarithm, 73

M

MAA, *see* Methacrylic acid
Magnesium carbonate, 36, 110
Magnesium oxide, 139
Malic acid, 189
    chemical structure, 190
Mandelic acid, 192
    chemical structure, 190
Medicines, time-released, 1–2
Meglumine ion, 8
Melting point, fusible alloys, 85–86
Menthofuran, 198
    chemical structure, 199
l-Menthol, chemical structure, 197
d-Menthone, 197
l-Menthone, 197
    chemical structure, 198
Menthyl acetate, chemical structure,
    197, 198
Mercaptans, 69, 71
Mercury, liquid
    freezing/boiling points, 126
    use in thermometers, 125–126
Metal ions
    in floor waxes, 96, 97
    and food spoilage, 41
Metalloanthocyanin, chemical structure, 65
Metals
    alkali, thermal ionization, 128
    combustible, 104
    fusable, 85–86
Methacrylic acid (MAA), 216
    chemical structure, 215
Methacryloxypropyl tris(trimethylsiloxy
    silane) (TRIS), 213
    chemical structure, 214
4-tert-butyl-4'-Methoxydibenzoylmethane, 202
    chemical structure, 205
Methoxyflurane, 10
    chemical structure, 12
Methyl α-cyanoacrylate, chemical
    structure, 130
3-Methylbutane-1-thiol, chemical
    structure, 70

Methyl chloroform, 82
Methyl methacrylate (MMA), chemical
    structure, 213
2-Methyl-2-propanethiol, 69
Mica, 121
Microencapsulation, 1–2, 76
Microspheres, 75
    in self-adhesive paper, 78
    silica, in opal, 167–168
Mineraloids, 167
Mirrors, 53
Miscibility, 93–94
Molality, 132
Molecular sieves (zeolites), 124
Molybdate, 142
Monochloramine, 95
Monomers, in contact lenses, 212
Montmorillonite clay, 123, 124
Myristic acid, 110

N

Natural gas, 70–71
Ninhydrin, chemical structure, 162
Nitrogen trichloride, 95
Nitroglycerine, 3
    chemical structure, 4
Nitrous oxide, 10, 14
    chemical structure, 11
Nonionic surfactants, 207

O

Octyldimethyl PABA, chemical
    structure, 203
2-Octyl-1-dodecanol, 193
    chemical structure, 194
Octyl methoxycinnamate, 202
    chemical structure, 204
Oleic acid, 110
Oleophobicity, 171
Opals
    amorphous structure of, 167
    iridescence of, 166
Optical brighteners, 180
Optical isomerism, 198
Optically variable ink, in currency, 158
Organ pipes, 144–145
Osmosis, 7
Osmotoxicity, 8
Ovoglobulin, 48
Ovomucin, 48

Oxazole, 180
    chemical structure, 182
Oxidation–reduction
    chlorine, 29
    in discoloration of pigments, 142–143
    hydrogen peroxide, 106
    indigo, 185
    iron and iron alloys, 140
    magnesium, 104
    in metal corrosion, 100
    ninhydrin, 163
    in solid rocket boosters, 56
    thiols, 69
Oxidation state, 105, 106
Oxybenzone, chemical structure, 202
Oxygen permeability, of contact lenses,
    212–213
Ozone, 201

P

PABA, 202
    chemical structure, 203
Padimate O, *see* Octyldimethyl PABA
Palmitic acid, 110
Panthenol, 193
    chemical structure, 194
Paper manufacture, 152–153
Partial pressure, 45
Patina, on copper, 100
Pearls, sold emulsion nature of, 167
Pentane, 71
Peppermint oil, 197
Peroxide compounds, 104
Petroleum gas, 71
PGA, *see* Poly(glycolic acid)
pH, 73
    albumen, 48
    of drain cleaners, 113
    effects on iron oxidation, 140
    as factor in time-released medicines, 3
    indicators of, 83–84
    of pool water, and chlorination, 29
    soil, factors affecting, 64
Phase separation, 76
Phenols, 115
Phenoxyethanol, 193
    chemical structure, 194
Phosphoric acid, chemical structure, 111
Photochemical reactions, 29
Photons, 128

Pigments
  in artist's paint, 141–143
  pearlescent, 120–122
Piperine, 34
PLA, *see* Poly(lactic acid) (PLA)
PMMA, *see* Poly(methyl methacrylate)
Polarizability, hydrocarbons, 108
Polyacrylic acid, 79
Polycarbonates, 115
  synthesis, 117
Polydimethylsiloxane, 109
  chemical structure, 108
Polyester
  ester linkages forming, 156
  in security thread of currency, 155
Polyesters, 114
  ester linkage of, 115
  fibers, versus cotton fibers, absorbancy
    of, 177
Polyethylene, based fibers, 173
Poly(glycolic acid) (PGA), 17
  synthesis, 16
Polyhydric alcohol, 188
Poly(lactic acid) (PLA), 17
  synthesis, 16
Polymerization, 79–81
  cyanoacrylate polymer, 131
  free radical vinyl, 118
Polymers
  in biodegradable sutures, 17
  in biomaterials, 20
  in bullet-resistant fabrics, 172
  in contact lenses, 212
  cross-linking, 216
  in drug encapsulation, 2, 3
  in floor waxes, 96–97
  heterochain, 114
  in microencapsulation, 76, 77
  siloxane, 108
  in superglues, 129
  in thermoplastics, 114
  in water-repellent fabrics, 169
Poly(methyl methacrylate) (PMMA),
    115, 212
  synthesis, 118
Poly-*para*-phenylene terephthalamide, 173
Polyphosphates, 111
Polysiloxanes, 119
POPU, *see* Polyalkylene oxide
    polyurethaneurea

Potassium bitartrate, 51
Precipitates/precipitation, 98
  agents, 112
  lauric acid-calcium, 110
  in seashell formation, 62–63
Pressure-sensitive chemical release, 75
Propane, 71
1,2-Propanediol, *see* Propylene glycol
1,2,3-Propanetriol, *see* Glycerin
Propellants, in solid rocket boosters, 55
Propylene glycol, 193
  chemical structure, 147, 195
  hygroscopic nature, 148
Pulegone, chemical structure, 199
Pyrophosphoric acid, chemical
    structure, 111

Q

Quartz, 167
Quaternium-15, 207
  chemical structure, 208

R

Radiopaque media, 6
Raschig process, 96
Red lead, 142
Reducing agent, 140
Reflector material, 160
Refraction, light
  definition, 153
  in pearlescent pigments, 121–122
  by solutes in ice, 132
Rubrene, 26
  chemical structure, 27
Rust, 140

S

Salicylic acid, 189
  chemical structure, 191
Salt, table, 35–36
Schaefer, Vincent J., 58
Scopolamine, 3
  chemical structure, 4
Seashells, 62
Sepiomelanin, 63
Sequestrants, 41, 111, 112
Sevoflurane, 12
Silica, 166
Silica gel, 123
Silica glass, 91

Silicone oil, 108
Silicones, 119
Siloxanes, 108–109
 hydrophobic nature of, 119
Silver, Spencer, 78
Silvering process, 53
Silver iodide, in cloud seeding, 58
Skunk odor, 68–69
Smoke, 147
Soda-lime silicate glass, 91
Sodium bicarbonate, 37, 102, 210
Sodium borosilicate, 91
Sodium carbonate, 83, 102
Sodium chloride, 127
 crystal structure, 136, 137
Sodium hydrosulfite, 185
Sodium hydroxide, 84, 113, 185
Sodium hypochlorite, 29, 95
Sodium ion, 8
Sodium silicate, 112
Sodium vapor lamps, 128
Solar flares, 60, 61
Solid emulsions, 167
Solubility, 33
Solutes, depression of solvent freezing
 point by, 132
Solvents, 82
Sorenson, Soren P. L., 73
Specific gravity, 93
Standard reduction potential, 140–141
Static electricity, and dust buildup, 107
Stearic acid, 110
Stearyl alcohol, 193
Stilbene, chemical structure, 180
Sublimation, 87, 88
Sulisobenzone, chemical structure, 203
Sunblocks, versus sun screens, 201
Supercooled clouds, 58
Surface coatings, polymeric, 96
Surfaces
 and adhesion properties, 130
 frictional forces, 108
Surface tension, 46–47
Surfactants
 amphiphilicity of, 206
 polar headgroups, types of, 207
Sutures, 15

T

Tartaric acid, 48, 50–51
 chemical structure, 190

Temperature
 freezing, of water, 132
 and gas solubility, 44–45
 and light stick luminescence, 24–25
 melting, fusible alloys, 85–86
 ranges, and thermometer types, 126
 and tin stability, 144–145
 and transdermal delivery of
  medications, 3
 and vapor pressure, 49
 of water, 38–40
Terephthalic acid, 173
Thermal analysis, of opal structure,
 167–168
Thermal breakdown, microcapsules, 76
Thermal coefficients of expansion, 93
Thermal conductivity, 54
Thermal decomposition, of limestone, 136
Thermal decomposition reaction, 102
Thermal expansion, coefficient of, 126
Thermometers, 125–127
Thermoplastic resins, 173
Thermoset plastics (thermoplastics), 114
Thief detection powder, 161–162
Thiols, 69, 71
Thymolphthalein, chemical structure, 83
Tin, 140
 melting point, 85
 stability of, 144–145
 standard reduction potential, 141
Titanium dioxide, 82, 121
 as paper coating, 153
 in sunblocks, 201
α-Tocopherol, 193
 chemical structure, 194
Transdermal delivery, medications, 3
Transitional metal complex ion, 97
Triazole, 180
 chemical structure, 181
1,1,1-Trichloroethane, 82
Trichloroethylene, 10
 chemical structure, 11
Tridymite, 167
Triethanolamine, chemical structure, 196
Triethanolammonium, 207
 chemical structure, 208
Triethylene glycol
 chemical structure, 147
 hygroscopic nature, 148
Trimethylchlorosilane vapors, 170

TRIS, *see* Methacryloxypropyl
  tris(trimethylsiloxy silane)
Tungsten, 87
  melting point, 88
TWEEN 20 polysorbate 20, 207
  chemical structure, 208

U

Ultraviolet light, classifications of, 201

V

Vaporization, enthalpy/entropy of, 38–39
Vapor pressure, 49
  of ethanol versus water, 188
Vinegar, 98, 99
  in laundering of denim, 185
N-Vinyl-2-pyrrolidinone (NVP), chemical
  structure, 215
Violanthrone, 26
  chemical structure, 27
Viscosity, 93
Vitamin E, *see* α-Tocopherol
Volatility, 12
Volume expansion coefficient, 126

W

Walker, Edward Craven, 92
Washing soda, 112
Water
  bound in opal, 167–168
  content of contact lenses, 213
  hardness of, 109
  heat capacity of, 68
  polarity of, 177
  vapor pressure of, 38–40, 188
Wurtzite crystals structure, 59

X

X-rays, in CAT imaging, 5–6

Z

Zeolites, synthetic, 123, 124
Zinc, as transitional metal complex ion, 97
Zinc oxide, in sunblocks, 201
Zingerone, 34
Zirconium, 139
Zwitterionic surfactants, 207–208